Advanced Security and Safeguarding in the Nuclear Power Industry

Advanced Security and Safeguarding in the Nuclear Power Industry

State of the art and future challenges

Edited by

Victor Nian, PhD

Senior Research Fellow
Energy Studies Institute
National University of Singapore
Singapore

ELSEVIER

ACADEMIC PRESS

An imprint of Elsevier

Academic Press is an imprint of Elsevier
125 London Wall, London EC2Y 5AS, United Kingdom
525 B Street, Suite 1650, San Diego, CA 92101, United States
50 Hampshire Street, 5th Floor, Cambridge, MA 02139, United States
The Boulevard, Langford Lane, Kidlington, Oxford OX5 1GB, United Kingdom

Library of Congress Cataloging-in-Publication Data
A catalog record for this book is available from the Library of Congress

British Library Cataloguing-in-Publication Data
A catalogue record for this book is available from the British Library

ISBN: 978-0-12-818256-7

For information on all Academic Press publications visit our website at
https://www.elsevier.com/books-and-journals

Publisher: Candice Janco
Acquisitions Editor: Maria Convey
Editorial Project Manager: Chiara Giglio
Production Project Manager: Kiruthika Govindaraju
Cover Designer: Matthew Limbert

Typeset by TNQ Technologies

Contents

Contributors... xi

About the Editor.. xiii

Preface xv

Acknowledgment.. xvii

CHAPTER 1   Recent advances in nuclear power technologies..... 1
Dmitrii Samokhin, PhD
1. Introduction 1
2. Methods for converting the energy of division to useful work..... 2
  2.1 Efficiency.. 2
  2.2 Heat transformation in electricity through mechanical
       work... 3
  2.3 Direct conversion of heat into electricity.......................... 4
  2.4 Other methods for converting fission energy into
       useful work.. 9
3. Materials for nuclear reactors: classification of nuclear
    reactors 13
  3.1 Types of fuel element structures 13
  3.2 Fuel.. 14
  3.3 Construction materials.. 18
  3.4 Coolant.. 19
  3.5 Moderator.. 22
  3.6 Absorber 23
  3.7 Classification of nuclear reactors 26
4. Conclusion... 28
Suggested reading.. 28

**CHAPTER 2   Non-power applications—new missions for nuclear
                energy to be delivered safely and securely.......... 29**
William J. Nuttall, PhD, BSc and
Gareth B. Neighbour, PhD, BSc
1. Introduction 29
2. New opportunities for civil nuclear energy 29
  2.1 Historical background.. 29
  2.2 Looking ahead 32
3. Security and safeguards and safety—the three S's................... 38
  3.1 Security 38
  3.2 Safeguards—nuclear nonproliferation risks...................... 39
  3.3 Safety—severe nuclear accidents................................... 41

    4. Summary remarks 44

Acknowledgments..44

References..44

CHAPTER 3   **Radiation hazards from the nuclear fuel cycle 49**
        Man-Sung Yim, ScD, PhD, MS, SM

    1. Basic concepts of radiation hazards........................49

    2. Overview of the nuclear fuel cycle..........................53

        2.1 Mining..54

        2.2 Milling...55

        2.3 Conversion...55

        2.4 Enrichment 56

        2.5 Fuel fabrication..56

        2.6 Reactor operation...56

        2.7 Storage of spent fuel...57

        2.8 Transportation of spent fuel 58

        2.9 Reprocessing of spent fuel.................................59

        2.10 Permanent disposal of spent fuel.....................59

        2.11 Decommissioning of nuclear fuel cycle facilities 60

    3. Presence or release of radioactive materials in the nuclear
       fuel cycle..61

        3.1 Mining.. 63

        3.2 Milling.. 66

        3.3 Conversion, enrichment, and fuel fabrication 67

        3.4 Nuclear reactor operation 67

        3.5 Reprocessing.. 72

    4. Concern with spent nuclear fuel.............................73

    5. Radiation concern in nuclear power plant decommissioning 77

    6. Conclusions...78

    References..79

CHAPTER 4   **Health effects of exposure to ionizing radiation 81**
        Wilner Martinez-López, MD, PhD and
        Manoor Prakash Hande, PhD, MPH

    1. Introduction..81

    2. Basic concepts of ionizing radiation 82

    3. Biological effects of ionizing radiation.....................83

    4. Acute effects of ionizing radiation..........................83

        4.1 Main overexposed cases to ionizing radiation 84

    5. Effects of low doses of ionizing radiation..................86

        5.1 Radiation workers and patients exposed to ionizing
            radiation.. 86

        5.2 Environmental radiation exposures..................... 87

6. Radiation-induced chromosome alterations 87
7. Transgenerational effects of radiation............................... 90
8. Epigenetics changes induced by ionizing radiations............... 90
9. Radiation protection... 91
10. Final remarks... 91
Acknowledgments.. 92
References... 92

**CHAPTER 5   Nuclear plant severe accidents: challenges and
prevention 99**
Wison Luangdilok, PhD, MS, BS and Peng Xu

1. Introduction 99
2. Fukushima Daiichi accidents and insights from the accident
analyses... 100
2.1 Event progression at unit 1 101
2.2 Event progression at unit 3 102
2.3 Event progression at unit 2 104
2.4 Explosions.. 105
2.5 Unit 1 explosion and spent fuel pool conditions 107
2.6 Unit 3 explosion and spent fuel pool conditions 108
2.7 Unit 4 explosion and spent fuel pool conditions 108
2.8 Remarks on the accident causes 109
3. Investigative studies of lessons learned 110
3.1 Causes of the fukushima accident as found by the US
National Academy of Sciences study............................ 111
3.2 Lessons learned from the National Academy of Sciences
Study.. 112
4. The new post-Fukushima regulatory body structure in Japan.... 115
4.1 Recommendations by the National Diet Independent
Investigation Commission 116
4.2 Recommendations by the Government Investigation
Committee.. 117
5. Post-Fukushima regulations and technology development........ 117
5.1 Japan's new regulatory regime 118
5.2 New regulatory requirements in Japan 120
5.3 Post-Fukushima regulatory requirements in the United
States.. 125
5.4 Development of Accident Tolerant Fuel Technology as
Risk Mitigation 128
6. Concluding remarks.. 131
Acknowledgment.. 132
References... 132

CHAPTER 6   Nuclear off-site emergency preparedness and response: key concepts and international normative principles.................................... 135
Günther Handl

1. Introduction 135
2. The international normative setting for emergency preparedness and response..139
　2.1 The IAEA-centered regulatory framework......................139
　2.2 Other international, regional, and industry-inspired nuclear emergency preparedness and response efforts........146
3. Specific emergency preparedness and response policy challenges and international regulatory responses 149
　3.1 Cross-border coordination of emergency preparedness and response: shared understandings and mutual trust.......149
　3.2 Emergency planning zones 152
　3.3 Event reporting and information sharing 157
　3.4 Validation of national emergency preparedness and response through international peer review and emergency exercises...162
4. Nuclear emergency assistance: global, regional, and bilateral arrangements...165
5. Conclusions...167
References...168

CHAPTER 7   International conventions and legal frameworks on nuclear safety, security, and safeguards........ 175
Paul Murphy and Ira Martina Drupady

1. Part 1: what is meant by nuclear safety, security, and safeguards? 176
2. Part 2: what are the key international conventions on safety, security, and safeguards?...180
　2.1 Introduction 180
　2.2 Safety...180
　2.3 Security 185
　2.4 Safeguards...188
3. Part 3: what are the additional conventions of relevance?........193
　3.1 Convention of environmental impact assessment in a transboundary context (the "Espoo Convention")..............194
　3.2 Convention on access to information, public participation in decision-making and access to justice in environmental matters (the "Aarhus Convention").................................196

4. Part 4: what are the significance of these regimes? –
reputational risk analysis 198
   4.1 Reputational risk 199
5. Part 5: concluding thoughts 205
References 205

CHAPTER 8   **Civil liability in the event of a severe nuclear disaster** **209**

Jonathan Bellamy, C.Arb

1. Introduction 209
   1.1 Legal issues 209
2. Nuclear new build programs 210
   2.1 Volume 210
   2.2 Geographic distribution 210
   2.3 New entrants 211
3. The international nuclear liability regimes 211
   3.1 Origins 211
   3.2 Regimes 213
   3.3 Key legal principles 214
   3.4 Convention on supplementary compensation for
      nuclear damage 215
   3.5 Nonsovereign extralegal initiatives 219
   3.6 The advantages of a harmonized international legal
      regime 220
4. Nuclear new build countries—national legal regimes 220
   4.1 The United States 220
   4.2 China 221
   4.3 Russia 222
   4.4 India 222
   4.5 The United Kingdom 223
   4.6 The United Arab Emirates 224
   4.7 The advantages of international convergence 225
5. Conclusion 225
References 226

CHAPTER 9   **Future challenges in safety, security, and safeguards** **229**

Tatsujiro Suzuki, MS

1. Introduction 229
2. Overview of global nuclear industry 229
3. Loss of public trust in nuclear safety 230

4. Challenges for nuclear security...231
5. Challenges for nuclear safeguards.....................................235
6. "Safeguarding" public from nuclear risks 238
7. Conclusion...239
References...240

Index...243

Contributors

Jonathan Bellamy, C.Arb
Barrister and Chartered Arbitrator, London, United Kingdom

Ira Martina Drupady
Research Associate, Energy Studies Institute, National University of Singapore, Singapore

Manoor Prakash Hande, PhD, MPH
Department of Physiology, Yong Loo Lin School of Medicine and Tembusu College, National University of Singapore, Singapore

Günther Handl
Eberhard Deutsch Professor of Public International Law, Tulane University Law School, New Orleans, LA, United States

Wison Luangdilok, PhD, MS, BS
President, H2Technology LLC, Westmont, IL, United States; Fauske & Associates LLC, Burr Ridge, IL, United States

Wilner Martinez-López, MD, PhD
Epigenetics and Genomic Instability Laboratory, Instituto de Investigaciones Biológicas Clemente Estable, Montevideo, Uruguay; Associate Unit on Genomic Stability, Faculty of Medicine, University of the Republic (UdelaR), Montevideo, Uruguay

Paul Murphy
Managing Director, Murphy Energy & Infrastructure Consulting, LLC, Washington, DC, United States

Gareth B. Neighbour, PhD, BSc
School of Engineering and Innovation, The Open University, Milton Keynes, United Kingdom

William J. Nuttall, PhD, BSc
School of Engineering and Innovation, The Open University, Milton Keynes, United Kingdom

Dmitrii Samokhin, PhD
Head, Department of Nuclear Physics and Engineering, Obninsk Institute for Nuclear Power Engineering of the National Research Nuclear University, Obninsk, Russian Federation

Tatsujiro Suzuki, MS
Professor, Doctor, Research Center for Nuclear Weapons Abolition, Nagasaki University, Nagasaki, Japan

Peng Xu
Idaho National Laboratory, Idaho Falls, ID, United States

Man-Sung Yim, ScD, PhD, MS, SM
Professor, Nuclear and Quantum Engineering, Korea Advanced Institute of Science and Technology, Daejeon, South Korea

About the Editor

Victor Nian

Dr. Victor Nian is a Senior Research Fellow at the Energy Studies Institute, National University of Singapore. Dr. Nian holds a PhD in Mechanical Engineering and a Bachelor in Electrical Engineering with a Minor in Management of Technology, all from the National University of Singapore. His expertise is in energy and nuclear policy, energy systems analysis, technology assessment, and integrated solution development. His research portfolio covers a diverse range of interdisciplinary projects supported by government agencies and private sector. In the spirit of "research and innovation without borders", he established UNiLAB on Integrated Systems Analysis Tools, which hosts a research network of more than fifteen academic and research organisations from around the world. He is a Founding Member and elected Council Member of the International Society for Energy Transition Studies amongst thirty-one other renowned individuals from organisations such as the United Nations Economic and Social Commission for Asia and the Pacific, Asian Development Bank and Economic Research Institute for East Asia. Dr. Nian was previously Visiting Fellow at the Hughes Hall, University of Cambridge and elected President of the Engineering Alumni Singapore, the official alumni society for NUS established in 1972.

Preface

Any sufficiently advanced technology is indistinguishable from magic.

Arthur C. Clarke

From the advent of the steam engine to the development of mass aviation, technology has been a remarkable boon to contemporary society. However, while technological advances have undeniably created tremendous opportunities on an unprecedented scale, they have also been fraught with myriad anxieties over their real and perceived dangers.

The history of nuclear energy is perhaps the epitome of the Janus-faced character of modern society's relationship with its growing technological mastery over nature. In 1904, the British physicist Ernest Rutherford wrote "If it were ever possible to control at will the rate of disintegration of the radio elements, an enormous amount of energy could be obtained from a small amount of matter." It was not till 35 years later that Albert Einstein's special theory of relativity—as emblematized in his famous formula "$E=MC^2$" for mass-energy equivalence—was proven, unlocking the floodgates to the tremendous energy potential of the fundamental building blocks of the universe—the atoms. Nonetheless, the allure of nuclear power has been repeatedly tempered from the long-lasting trauma of its birth in the flames of the Hiroshima and Nagasaki atomic bombings, to the public backlash incited by the ongoing impacts of the Fukushima Daiichi nuclear accident. It is therefore unsurprising that the "Three S's"—safety, security, and safeguards—have become the three fundamental pillars of the global nuclear power industry.

Our addiction to fossil fuels has caused devastating damage to the environment, biodiversity, and public health. Moreover, the costs of inaction are rapidly mounting as the unrelenting accumulation of anthropogenic greenhouse gases threatens to tip the damaging transformations ecosystems worldwide past the point of no return.

While the nuclear power industry will continue to face evolving technological and institutional challenges, national regulatory bodies and the International Atomic Energy Agency have promulgated policy, regulatory, and legal instruments as additional security measures to safeguard the public interest as well as the global nuclear power industry in delivering safe and reliable clean energy to our modern society. This book provides a comprehensive overview of these measures, and how they have been updated to address future challenges. As such, the book is particularly relevant to countries with an interest in developing a nuclear power industry but which are not yet a nuclear state, as well as countries where education to improve society's opinion on nuclear power is crucial to its future success in low carbon development.

Books on atomic energy have almost exclusively focused on the safety aspect. However, the equally important issues of security and safeguards have often been unduly overlooked. Motivated by the need to bring a balanced review of a wide-ranging scientific, engineering, policy, regulatory, and legal issues facing the nuclear power industry, we embarked on this endeavor to deliver a much needed review of the latest developments in "Three S's" as well as other related and important subject matters within and beyond the nuclear power industry. This book presents the state of the art in an accessible form for a wide-ranging audience that is suitable for nuclear industry practitioners, scientists, engineers, lawyers, educators, and policymakers.

Acknowledgment

We are most grateful of the contributing authors in delivering excellent reviews of important subject matters for this edited book. In addition, we would like to record our thanks to Philip Andrews-Speed, Yousry Azmy, S.K. Chou, and Egor Simonov (family name in alphabetical order) for generously providing assistance in various commendable ways. The editor would like to thank everyone on the editorial team. Special thanks to Michelle Fisher, the ever-patient Editorial Project Manager, for taking this project through to its success.

Victor Nian, PhD, BE
Senior Research Fellow, Energy Studies Institute,
National University of Singapore, Singapore

Recent advances in nuclear power technologies

Dmitrii Samokhin, PhD

Head, Department of Nuclear Physics and Engineering, Obninsk Institute for Nuclear Power Engineering of the National Research Nuclear University, Obninsk, Russian Federation

1. Introduction

This chapter is based on lectures that the author gave for the course "Construction of Nuclear Reactors" for a number of years to senior students at the Obninsk Institute for Nuclear Power Engineering of the National Research Nuclear University, Moscow Engineering Physics Institute.

Textbooks and manuals on this subject are obsolete or designed for students of other specialties. They generally contain descriptions of specific design solutions used in various nuclear reactors. However, they almost never provide guidance for developers in making these decisions, as well as the side (possibly nonphysical) factors affecting them. The material in this chapter is intended to fill this gap to the best possible extent.

Nuclear power reactors are mostly by-products of the defense industry (except maybe Canada deuterium uranium reactors). They were created by experts whose mindset was to solve major military tasks in which the possibility of human and material loss was considered natural and only to be decreased.

Therefore, when the first generation of nuclear reactors was created, their technical and economic quality was optimized without much careful consideration of security issues. Moreover, it was believed that even when the project violated safety requirements, if they were minor, "" the authorities into"" the project with the right influence. The bitter lessons of disasters such as the Three Mile Island nuclear power plant (NPP) in the United States, the Chernobyl NPP, and the Fukushima Daiichi NPP showed that neglected security issues can also lead to disastrous economic consequences, not just overexposure of radiation to personnel, the population, and the environment.

I hope that for a new generation of designers of nuclear reactors, the formation of the style of thinking that is one purposes in writing this chapter will result in more care and attention to safety in nuclear power.

It is assumed that the reader has the original information from the course "Nuclear and Neutron Physics" (i.e., the designation "U-235" denotes the isotope uranium element with a nucleus containing 235 nucleons).

Advanced Security and Safeguarding in the Nuclear Power Industry. https://doi.org/10.1016/B978-0-12-818256-7.00001-5

Considering that any textbook for convenience should assume a maximum degree of self-sufficiency and contain at least links to information from other publications, the author found it necessary to include in the guide some materials related to other courses but required to understand the key moments. This information is concentrated mainly in the first chapter, which can be considered as a first approximation, an introduction to the profession.

All of us, especially the students, are tired of the abundance of mathematical formulas, equations. and proofs in the educational and scientific literature in the disciplines of engineering and physics. Therefore, knowing that it is impossible to avoid them completely, the author has sought to use only the minimum number of them necessary.

The author would like to thank Professor Volkov Yuri Vasilievich (Obninsk Institute for Nuclear Power Engineering of the National Research Nuclear University Moscow Engineering Physics Institute), as the head of the material prepared on the basis of its eponymous manuals.

2. Methods for converting the energy of division to useful work

2.1 Efficiency

When you convert, a part of energy is lost. If a machine- or apparatus-implemented process changes and/or energy is transferred, the efficiency of this process is usually characterized by a coefficient of performance. The circuit of this device has the form shown in Fig. 1.1.

Efficiency is defined as the ratio of useful work to input energy: $\eta = E_{use}/E_{sup}$. The energy conservation law $E_{sup} = E_{use} + E_{loss}$ is then:

$$\eta = 1 - E_{loss}/E_{sup}.$$

It is known from the course on thermodynamics that to obtain work from heat continuously, it is necessary to have a working fluid that would carry out a sequence of circular processes (i.e., such processes in which it would periodically return to its

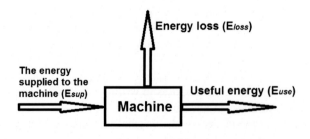

FIGURE 1.1

Machine scheme for the process of conversion and/or energy transfer.

original state). In each cyclic process, the working fluid receives a quantity of heat Q, the primary energy source (in our case, of a nuclear fuel) at a sufficiently high temperature and sends a minimal amount of heat Q_2 to the environment (water or air).

Because the working fluid after the cycle is returned to its original state and does not change its internal energy, in accordance with the law of conservation of energy, the difference in heat is converted into work:

$$L = Q_1 - Q_2.$$

The possibility and efficiency of heat conversion to other forms of energy (mechanical or electrical) are primarily determined by the temperature at which the Q_1 heat can be transferred to the working fluid. The temperature at which the heat given, Q_2, is also significant. Because warmth is given to the environment, in reality this temperature varies within a narrow range determined by fluctuations in the temperature of the environment.

The efficiency of converting heat into work is evaluated by the thermal efficiency:

$\eta_{t-max} = L/Q_2 = 1 - Q_2/Q_1$. From the course on thermodynamics, it is known that if the absolute temperature T_1 of heat supply Q_1 and the temperature T_2 of heat removal Q_2 are set, the maximum possible efficiency:

$$\eta_{t-max} = 1 - T_2/T_1$$

Such efficiency can theoretically be obtained by the so-called Carnot cycle, which in practice cannot be realized. All real cycles in which the highest heat input temperature T_1 and the lowest temperature of heat removal T_2 have a thermal efficiency:

$$\eta_t \leq 1 - T_2/T_1$$

The design of the nuclear reactor should be such that the temperature of the fuel, and accordingly, the coolant, is as high as possible. In this case, the efficiency of the reactor as a heat engine will be maximum.

2.2 Heat transformation in electricity through mechanical work

In all of the NPP, thermal energy obtained by nuclear fuel is converted into mechanical vapor while expanding in the turbine, which in turn rotates a generator that generates electricity. A simplified view of a steam power plant, is shown in Fig. 1.2. It includes (1) a heat source, (2) a steam turbine, (3) a condenser, and (4) a pump.

Steam power equipment works on the so-called Rankine cycle (i.e., the cycle in which the working fluid is at a high temperatures and steam works in the turbines and at low temperatures as liquid). Because the fluid is practically incompressible, pump 4, which serves to raise the pressure rise and circulate the working fluid, consumes relatively little energy, L_M. Maximum efficiency is:

$$\eta = 1 - (Q_2 + L_M)/Q_1$$

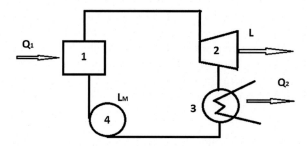

FIGURE 1.2

Simplified view of steam power plant.

Nuclear power may be offer options for transferring heat to the working body:

1) The heat source 1 (the reactor itself)
2) Heat source 1 (the heat exchanger to which heat is supplied to the reactor via an intermediate circuit (Fig. 1.3)
3) More intermediate heat exchangers (Fig. 1.4)

This traditional conversion of heat into electricity and its impact on the design features of the actual nuclear reactors are further discussed in detail in later chapters. The maximum efficiency that can be achieved in this scheme is 33%–40%.

2.3 Direct conversion of heat into electricity

Because the original form of energy in an energy conversion device is direct heat, the efficiency of obtaining electricity is subject to the restrictions of the second law of thermodynamics and cannot exceed the efficiency of the Carnot cycle for the same temperature interval.

There are two methods for direct conversion:

- thermoelectric and
- thermal electron emission.

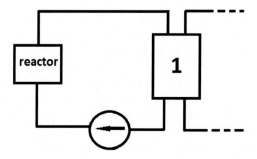

FIGURE 1.3

Dual nuclear power plant (NPP) scheme.

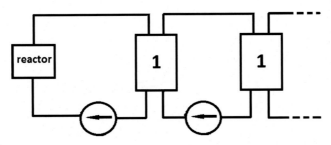

FIGURE 1.4
Three-loop-circuit nuclear power plant (NPP).

2.3.1 Thermal electric generators.

The work of thermoelectric generators (TEG) is based on the thermoelectric effect discovered in the last century: the Peltier and Seebeck effect.

We will consider the *Peltier effect.* If, after a junction of dissimilar conductors (metals and semiconductors) constant current I is skipped, the junction depends on the direction of current heat released or absorbed:

$$Qp = \alpha IT.$$

where α is the factor depending on the properties of selected conductors and T is the junction temperature.

We will consider the *Seebeck effect.* If the connection consists of two dissimilar conductors junctions at different temperatures, T_1, and T_2, an electromotive force (e.m.f.) E is proportional to the temperature difference:

$$E = \alpha(T_1 - T_2)$$

where α is coefficient of thermal e.m.f. or Seebeck coefficient.

Both effects complement each other and have the same physical nature, in that if any has free electrons, it tends to come to thermal equilibrium with the surrounding nuclei of the material. Therefore, in both formulas, the coefficient α is the same.

The diagram of one TEG is shown in Fig. 1.5. Thermoelectrodes 1 and 2 are made of different materials electrically connected to junctions A and B. Electrode 2 is broken and the gap is included (key 3) and load R.

If junctions A and B are kept at different temperatures, $T_1 > T_2$, the open circuit will be important in the difference (E). When the potentials close key 3, the circuit and a load current flows (I). However, according to the Peltier effect, when a current flows through junction I, dissimilar conductors at this junction are absorbed or released heat (Q_n). For example, in junction A, current flows from conductor 1 to conductor 2, and thus, it is absorbed heat, $Q_1 = \alpha IT_1$, which must fail. Then, in junction B, on the contrary, current flows from conductor 1 to conductor 2, whereby the junction generates heat $Q_2 = \alpha IT_2$, which must be removed.

When current I flows in the circuit, where e.m.f. effect E is the electric energy produced by $L_{el} = EI$, i.e.,

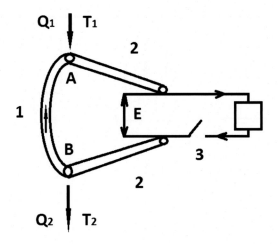

FIGURE 1.5

Thermoelectric generator.

$$L_{el} = \alpha(T_1 - T_2)I$$

in the ideal case

$$L_{el} = Q_1 - Q_2$$

for such an ideal TEG, efficiency would have been

$$\eta_{ideal} = L_{EL}/Q_1 = \alpha(T_1 - T_2)\, I/\alpha T_1\, I = 1 - T_2/T_1 = \eta_{t,\,max}$$

In this case, the efficiency is that of the Carnot cycle.

However, in reality, this efficiency cannot be obtained. Along with the processes described earlier for TEG, others substantially reduce efficiency. First, owing to the temperature difference between the junctions of the electrodes by 1 and 2, specific thermal conductivity from the hot junction to the cold heat results in flow Q_T. This heat is useless. It is at a constant L_{EL} increases the required heat Q_1 (i.e., it reduces efficiency). The amount of heat at a given Q_T difference T_1-T_2 is proportional to the coefficient of thermal conductivity λ and conductor cross-sectional area and inversely proportional to its length.

2.3.2 Thermoelectric generator accepted quality measure Q factor

$$z \sim \alpha^2/\lambda$$

The larger the z (i.e., the more the TEG performance), as measured by the coefficient α, and the less losses of heat is measured by thermal conductivity coefficient λ, the higher efficiency results for the TEG. This is an ideal to which we should strive to create TEG: it is necessary to provide better than the Q factor of $2 \cdot 10^{-3}$ materials can withstand, and the system must maintain the temperature of the hot junction ~ 1000 K.

The most successful materials for thermoelectrodes are considered to be alloys and compounds of elements of groups IV-VI of the periodic system: tin, lead, bismuth, antimony, tellurium, selenium, germanium, and silicon (semiconductor). The Q coefficient values for their z may reach $2 \cdot 10^{-3}$ to $3 \cdot 10^{-3}$ 1/degrees. The strong temperature dependence of z means that you can actually reach $1.5 \cdot 10^{-3}$ 1/degrees.

Typically, the TEG is a sequence of thermocouples connected in series special switching plates forming junctions. The result is a group of so-called hot junctions operating at a temperature T_1 and cold junctions operating at a temperature T_2 ($T_1 > T_2$) Fig. 1.6 is a diagram of the TEG. The full e.m.f. developed by the TEG is the sum of the individual elements of the e.m.f. The TEG circuit (terminals A and B) passes through the load and the switching thermoelectrode plate passes the same current.

As a result, the hot junctions absorb, emitting heat and cold. To maintain constant temperatures T_1 and T_2 to the hot junctions, it is necessary to sum Q_1 heat, and from cold to draw Q_2. TEG efficiency is less than a single element of the additional losses in the switching plates.

Owing to the high cost and low efficiency, TEG are used in large-scale stationary power generation. However, they are used widely in space solar energy. The energy source is nuclear reactors or radioisotope sources. Achievable electrical power is up to tens of kilowatts. The materials used are germanium-silicon alloys.

TEG is placed in a nuclear reactor and arranged to the supply heat removal within the constraints of mass and dimension is disadvantageous. Therefore, for cosmic power plants, TEG is delivered in refrigerator emitters. Hot junctions are usually at a temperature of $T_1 \sim 900$K, which provides the pumping of liquid metal (LM) coolant. The efficiency of such power plants is 5% or less.

2.3.3 Thermal electron energy converter

The basis of thermal electron energy converter (TEC) rest in the thermoelectric emission phenomenon: if any metal heated to a certain temperature, t, is placed in a vacuum, a certain amount of its electrons will move in vacuum. In this transition,

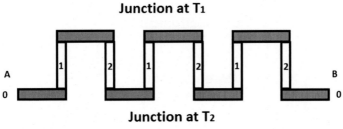

FIGURE 1.6

Thermoelectric circuit.

electrons must overcome an energy barrier called work output, φ, typically a few electron volts component.

At low temperatures, the average energy of the free electrons is substantially less for φ and only a tiny fraction of the electrons is emitted into a vacuum. The number of free electrons increases sharply with increasing metal temperature. The phenomenon of thermionic emission is widely used in electron tubes and electron accelerators.

The heated metal body is placed in a vacuum. After a while, when the electron cloud and the potential difference is set, further electron emission stops. Under these conditions, however many electrons come out of the metal, the same amount returns owing to natural condensation. The equilibrium potential difference between the metal and the electron cloud is equal to the metal φ output.

Electrons emitted by the body (cathode-emitter) can be selected, for example, by placing it close to the cathode the anode (collector) and applying a voltage of appropriate sign. The maximum quantity of electricity that can be selected in a unit of time is called the saturation current. The density of this current, i, can be calculated by the Richardson formula:

$$i = AT^2 \exp\left(-\varphi / KT\right).$$

in which $A \approx 120a/sm^2K^2$, the Richardson constant; φ is the metal work function; and k is Boltzmann's universal constant.

2.3.4 Where is the required voltage?

Voltage is applied to the cathode and anode from an extraneous electric source, which acts continuously, and the circuit is closed through a load. The current will flow through the circuit, determined by work function φ and temperature T of the cathode. All electronic lamps work this way. However, they are consumers rather than energy.

Work energy sources are organized differently. When placed in a vacuum, two electrodes of different metals with different work functions, φ1 and φ2, have some potential difference, Δφ (Fig. 1.7) established between them.

If electrodes 1 and 2 are the same temperature, no current will exist (otherwise it would be perpetual motion) for the closure chain. If electrode-emitter I has a higher

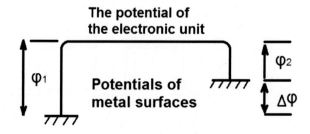

FIGURE 1.7

Potential of the electrode clouds.

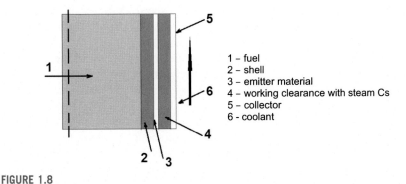

1 – fuel
2 – shell
3 – emitter material
4 – working clearance with steam Cs
5 – collector
6 - coolant

FIGURE 1.8

Thermoelectric generators in the form of a fuel rod. 1, fuel; 2, shell; 3, emitter material; 4, working clearance with steam Isotope Cs-133; 5, collector; 6, coolant.

temperature than the electrode-collector, the circuit closing electrons from the emitter to the collector will be used.

If the emitter temperature is not maintained, it is cooled, because the selection of the electrons' electrode cools (Edison effect). To keep the emitter temperature constant, heat must be supplied to it:

$$q = i(\varphi + 2kT/e)$$

per unit surface, where e is the electron charge; the rest of the notation is as explained earlier. When electrons are from the vacuum in the manifold, an appropriate amount of heat (heat of condensation) is allocated; to maintain a constant temperature of the collector, this heat must be removed.

In the NPP, which is based on this principle of energy conversion, it is possible to create a compact reactor-converter (RC), in which all energy-producing parts are built into the core and contain no moving parts. On the outside, there is only a cooling circuit. In this scheme, the fuel element itself is designed as a TEC (Fig. 1.8). Such designs have been created and work successfully: for example, the domestic space NPP with RC "Topaz". Creation of such a plant is an engineering challenge because TEC has to work at high temperatures, higher currents, and neutron fluxes. This is particularly frustrating because the properties of materials under irradiation can vary greatly:

The possibility is considered of creating a combined NPP direct conversion and power conversion machine method.

2.4 Other methods for converting fission energy into useful work

2.4.1 Magneto-hydrodynamic method

The operating principle of a magneto-hydrodynamic (MHD) generator is substantially identical to the usual operation principle of the electromechanical generator. Just as in ordinary e.m.f. in the MHD generator, it is generated in the conductor, which crosses the magnetic field lines at a certain speed. However, if conventional generators' movable conductors are made of solid metal in the MHD generator, they represent a flow of conductive liquid or gas (plasma).

The working fluid enters generally rectangular conductive flow channel at velocity W. In accordance with the laws of electrodynamics, when the working fluid moves in a magnetic field with induction B, an electric field with intensity E is induced in it. At the same time, an electric driving force arises on the electrode walls. It is equal to the product $E * b$, where b is the transverse dimension of the channel. When electrodes are attached to the outer Rn load and the working fluid is current (I), this current flows in the channel and interacts with the magnetic field, in which each volume of the working fluid acts against the electromagnetic directional motion retarding stream. In this way, the kinetic energy of the working fluid flow is ultimately converted into the energy of the electric current.

There are plasma and liquid metal MHD installations. These installations are divided according to the type of working fluid. Furthermore, these settings can be open and closed loop. Open cycle plants use the working fluid only once and are treated as add-ons to conventional steam power plants running on fossil fuel. In MHD facilities, closed cycle working fluid undergoes cyclic changes, repeatedly passing the working volume of the MHD generator. These settings may use a nuclear reactor as a heat source.

There are three ways to use nuclear reactors in MHD facilities.

1. Fast neutron reactors use liquid metals as a coolant. Liquid metals (LM) are capable of conducting electric current and they are ideal working fluids for magnetohydrodynamic (MHD) generators. However, the difficulty is that it takes more velocity, w, which in plasma MHD generators is obtained by expansion in the nozzle. LM is incompressible and dispersion through expansion is impossible. If pump disperses, it will take more energy than what is produced in the end. Some techniques overcome this disadvantage: part (up to 12%–13%) evaporated LM (injection method); and added (30%) gas (two-phase flow diagram). Both methods make the working fluid compressible in an MHD generator.

2. Gas-cooled reactor. The working fluid is a plasma consisting of inert gas. Plasma is obtained by heating it in a nuclear reactor. In this case, the problem is obtaining very high temperatures of working fluid and at the same time ensuring the stability of the reactor materials.

3. The gas-phase nuclear reactor. The working fluid is a fissile material in a gaseous state. It is pumped through the reactor so that the release owing to fission heat is carried away by the flow of the working fluid. Unlike conventional nuclear reactors, in conventional solid fuel, heat from it is taken away in liquid or gaseous coolant.

In the third scheme, the gas-phase reactor, it is possible to raise the temperature of the working fluid to 10,000 K values or greater. At this temperature, the fissile material is ionized into a plasma without requiring an ionizing additive.

The working medium in the gas phase reactor can be the vaporized uranium or any sufficiently volatile compound, such as uranium hexafluoride (UF_6), subliming

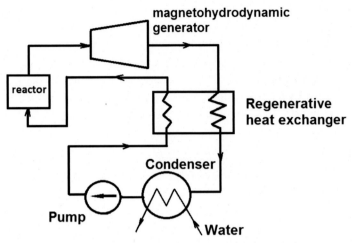

FIGURE 1.9

Magneto-hydrodynamic installation with gas-phase nuclear reactor.

at $T = 53°C$ and atmospheric pressure. This UF_6 property allows MHD installation to implement the same thermodynamic cycle as in a conventional steam power plant. A possible schemes for an MHD plant with a gas-phase nuclear reactor is shown in Fig. 1.9.

In the condenser, the water evaporates and is involved in a conventional steam power cycle, giving energy to the turbine. The regenerative heat exchanger, UF_6, extending into the reactor is heated to the desired temperature as a result of the high-grade heat UF_6, spent in the MHD generator.

2.4.2 Nuclear rocket engine

Nuclear rocket engines (NRE) are directly converted to heat produced by the fission of heavy nuclei in the nuclear reactor, the kinetic energy of movement of a rocket. When organizing manned expeditions to planets of the solar system, humanity will inevitably come replace chemical rocket engines (HRE) with NRE. Only NRE can provide the necessary motion parameters that guarantee a return expedition to Earth.

Understandably, high-quality NRE can be ensured only if the ground nuclear power is of sufficiently high development. This inspires optimism for the development of domestic nuclear energy, because by its nature, humanity must strive to reach at least the closest planet to Earth.

Using only the simplest information from the general physics and theory of rockets, we will compare the HRE and NRE. The HRE working body itself is a source of energy, which then can be used in rocket motion energy. Energy arises during the combustion of chemical fuels and a corresponding increase in combustion temperature compared with the initial fuel temperatures and oxidizer. The composition of the working fluid of these engines must include an oxidizing agent, preferably as efficient as possible, such as oxygen or fluorine. The means that:

1) except in rocket fuel, it is necessary to carry the same large extra load of oxidant;
2) the carriage-efficient oxidant in large quantities is dangerous, which is especially important for manned missions.

Because the rockets move in accordance with the law of conservation of momentum, it is necessary for the working fluid to be ejected from the rocket at the greatest possible speed. Accordingly, the most important characteristic is the quality of the rocket thrust:

$$R = m \, v.$$

where m is mass flow rate of the working fluid and v is exhaust velocity from the nozzle. It is known that the velocity of the gas at the outlet of the nozzle is an ideal:

$$V \sim \frac{\sqrt{T_k}}{M}$$

where T_k is the gas temperature upstream of the nozzle and M is the relative molecular mass of the flowing gases.

From these equations, immediately visible NRE advantages may be compared with HRE.

1. The maximum speed that can be achieved in the HRE is:

$$V \leq 6.5 \, \text{km/s}$$

The atomic mass of the working fluid cannot be less than 18 in a chemical rocket engine. This is due to the presence of oxidizing atoms in the working fluid. The NRE may be one of hydrogen working fluid from $M = 1$. From the latter equation, it follows that because of this (i.e., at the same T_k), speed expiration can be increased to about 18 times (i.e., three to four times). An increase of possibly even more increases this advantage.

2. Because of the heat source of the NRE (a nuclear reactor, and thus no chemical burning or burning of fissile nuclei), there is no need to have an onboard oxidizer. The mass and dimensions of the engine and its systems are noticeably reduced.

The huge energy consumption of nuclear fuel allows one to realize these benefits. Fig. 1.10 shows a typical circuit NRE.

Fig. 1.10 requires only one comment: supply the working fluid (indicated by arrows) arranged through the wall of the nozzle in order to, firstly, to cool them, and secondly to utilize the heat inevitably emanating from the nozzle wall, for increasing efficiency by preheating the working fluid prior to entering the reactor.

This chapter discusses how to convert fission energy into useful work. These methods are either already mastered (steam power, thermoelectric, thermionic) or have a good implementation prospect (magnetohydrodynamic generator, nuclear

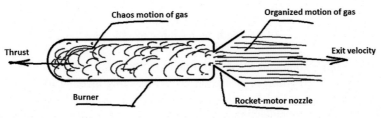

FIGURE 1.10

Scheme of nuclear rocket engine.

rocket engines). This perspective is prepared by developed modern technologies and scientific and technical potential of advanced countries.

3. Materials for nuclear reactors: classification of nuclear reactors

Because fuel is concentrated in fuel rods, it is the main element of a nuclear reactor core. Only by knowing the type design of the fuel elements is it possible to understand what materials are best to produce.

3.1 Types of fuel element structures

The main structural elements of a typical fuel element are depicted in Fig. 1.11.

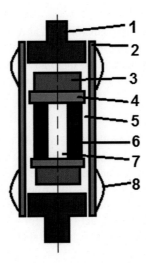

FIGURE 1.11

Typical fuel element structure. 1, End part of fuel rod; 2, Shell; 3, Shielding volume; 4, Separation element; 5, Gas gap; 6, Fuel core; 7, Central hole; 8, Distance element.

Comments on Fig. 1.11:

1. The active volume is filled with fuel with a specific enrichment of fuel (0.7% −90%).
2. The screening volume is filled with uranium-238 or a reflector.
3. The gap is filled with gas or a good heat-conducting material to provide large heat fluxes.
4. The gas volume for collecting gaseous fission fragments and the holding pressure in a fuel rod are within the allowable range (but not always).
5. In the case of the production of the screen and fuel from corrosion-interacting materials, separation elements are placed between them.
6. The central opening is used to collect gaseous fission fragments and compensate for possible fuel swelling.
7. The distance element is designed to provide the required flow cross-section between adjacent fuel rods.
8. The end pieces are used to secure the fuel rods in the fuel assembly.

3.1.1 Types of fuel rods

1. Cylindrical:
   a) It is located in a separate cylindrical channel (fuel rods).
   b) If using a lot of fuel in the same channel, it is called rod, and the whole structure is a fuel assembly.
2. Plate (Fig. 1.12A).
3. Ring (Fig. 1.12B), washed by the coolant from both sides.
4. The tube (Fig. 1.12C), washed by the coolant only from the inside.
5. The ball (Fig. 1.12D).

It is advisable to describe materials for nuclear reactors from the place of energy release to the periphery. This will always call attention to the merits of the materials and their defects. Nature is structured so that there is nothing ideal. Therefore, there are no materials ideally suited in all respects to nuclear reactors.

3.2 Fuel

Fuel is operated under severe conditions, exposure to high temperatures, variable pressures, and thermal stresses owing to differential thermal fields. If you do not comply with normal operating conditions, there may be corrosion and erosion. Therefore, the fuel rods work under stressful circumstances and the process of their mechanical properties changes. Because the fuel is irradiated, after a while, it is no longer possible to work in direct contact (i.e., if for some reason the fuel rod ceases to fulfill the required conditions, such as that the shell has lost tightness, it cannot be repaired.

An important effect for the design of fuel rods in the fuel is radiation swelling. Because of isolation of the fuel fission gases (e.g., xenon), it swells (changes geometry). This effect largely determines the allowable fuel burn-up.

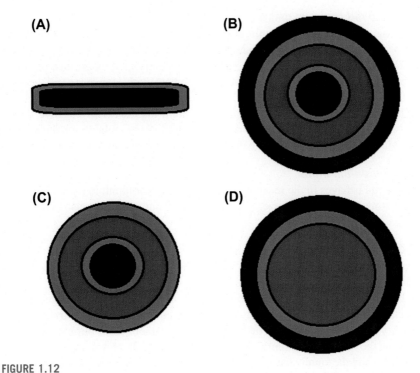

FIGURE 1.12

Types of fuel elements.

When you create nuclear reactors to determine the optimal design of the fuel elements and suitable materials, spend a lot of experimentation, which will result in an analysis of selected materials and fuel modes of operation. From the viewpoint of the designer (constructor and calculators), the main limiting parameters for fuel are:

1) permissible maximum fuel temperature
2) maximum allowable burn-up.

3.2.1 Pure uranium metal

According to the neutron-physical point of view, the best material for fuel could be considered pure uranium metal impurity nuclei with no other elements. In this case, it is necessary to take into account the positive and negative properties of fissile material. However, pure uranium metal is an unstable material with inclined formability at high temperatures.

The maximum temperature at which it can still work without significant swelling and forming is 400°C. For example, when T = 500°C, swelling reaches 20%.

The maximum burnup can be no more than 1% of fissile nuclei in nuclear fuel.

The pure uranium metal-corrosion active material reacts with almost all materials used in the reactors to 400°C. For reasons outlined earlier, pure uranium metal may be used in reactors where fuel T is less than 300–400°C.

3.2.2 Alloyed uranium metal

To improve the properties of uranium metal, it is typically alloyed with molybdenum (up to 10%). The more molybdenum, the higher limit temperature (up to 650−715°C) is the burn limit (2.5%−2%). Corrosion properties are also improved. For example, the allowable temperature in contact with stainless steel becomes 550°C. This is a positive property.

There are also drawbacks. Doping of uranium also leads to negative effects, because for molybdenum, parasitic neutron capture occurs in the thermal energy region, which worsens the neutron balance, especially in thermal neutron reactors.

3.2.3 Plutonium metal

The material behaves in the reactor about the same as for uranium metal. Moreover, its operating limit is even lower. Therefore, plutonium metal is also subjected to doping.

3.2.4 Uranium oxides and plutonium

Dioxides are used in nuclear reactors as fuel: UO_2 and PuO_2. The basic parameters of these materials are shown in Table 1.1.

The disadvantages of these materials in some ways are their advantages in other respects. For example, the relatively low density and thermal conductivity make it necessary to reduce the size of fuel rods to prevent them from burnout caused by exceeding the limit temperature at the center of the fuel rod. However, the "porosity" of dioxides and the small size of the fuel elements can achieve relatively high burnouts, to 10% and above.

The low density and the presence of O_2 are bad for fast reactors; they soften the neutron spectrum of the core and increase the resolution.

3.2.5 Carbide uranium and plutonium

The most common monocarbides are UC and PuC. The density is higher than for dioxides and is ~ 14 g/cm^3. The thermal conductivity is close to that of uranium metal and is about 15−20 kcal/(m.h°C) (i.e., ~ 10 times higher than that of UO_2). The melting temperature is also high ($\sim 2450°C$), which allows the fuel to be used at temperatures up to $\sim 2200°C$.

The main disadvantage is the shift of the neutron spectrum to the region of thermal energies in a fast reactor due to the presence of carbon nuclei.

3.2.6 Dispersion fuel

This is a heterogeneous mixture, in which the fuel phase is (uniformly) distributed in the non-fuel matrix material. In this case, each nuclear fuel particle is a kind of micro fuel rod enclosed in a shell, which performs the role of the matrix.

Metallic uranium and plutonium, and their alloys, as well as various compounds, are used as a fuel phase. Metals, ceramic materials, graphite are used as a fuel matrix.

Table 1.1 Basic parameters of uranium dioxide and plutonium.

Material	Melting, T°C	Permissible, T°C	Density theoretical, g/cm³	Density achievable, g/cm³	Thermal conductivity, kk/ (m·h·°C)
UO_2	~2800	~2500	~11	~8–10	1.5–2
PuO_2	~2200	~2000	~11	~8–10	1.5–2

3.3 Construction materials

A critical part directly adjacent to the fuel is the fuel cladding. Fuel cladding works under the difficult conditions of high temperature and active media. The most important harmful factor is that all kinds of radiation occur in the reactor, pass through the shell, or from the fuel out of the fuel element or the outside the fuel element to the fuel. In addition, coatings may deteriorate upon exposure to spacing and other nodes that contact fuel rods, causing corrosion and erosion (washout) to the heating medium and from the fuel (e.g., under the influence of fission products).

On the outer surfaces of the shells corrosion products can deposit, as can other reactor elements.

Basic requirements for materials skins are:

1. Low neutron-capture cross-section in the operating range of the reactor.
2. Corrosion and erosion resistance to the coolant at given parameters, and compatibility with fuel and fission products.
3. High thermal conductivity.
4. Satisfactory mechanical properties (strength, ductility, and creep) with radiation exposure, causing changes in these properties.
5. Manufacturability (i.e., possibility of manufacturing pipes and other profiles required, weldability).
6. Cost-effective and affordable.

Options main structural materials are listed in Table 1.2.

3.3.1 Aluminum and its alloys

Aluminum and its alloys have high thermal conductivity and a low thermal neutron capture cross-section. It can be used at temperatures up to 300°C. Therefore, it is used in low-temperature (for example, research, pool-type) reactors, including those in contact with uranium metal.

3.3.2 Magnesium and its alloys

Magnesium and its alloys have a low neutron absorption cross-section and high thermal conductivity and are cheap and available. For example, the Magnox alloy

Table 1.2 Parameters of main structural materials.

Material	Density g/cm^3	Melting temperature, °C	Coefficient of thermal conductivity. W/(m °C)	Cross-section absorbed barn
Al	2.7	660	210	0.215
Mg	1.74	651	159	0.059
Zr	6.5	1845	23.9	0.180
Stainless steel	7.95	~1400	14.6	2.880
Graphite	1.65	~3650	130–170	0.0045

(doping $\sim 0.6\%$ Zr, 0.8% Al) has good mechanical and corrosion properties. However, magnesium and its alloys are unstable in water. Therefore, do not use these in water reactors. The main application of Magnox is in gas-cooled reactors, in which the heat carrier is CO_2 (English and French reactors with metallic uranium as fuel). It is applied at temperatures up to 400°C.

3.3.3 Zirconium and its alloys

Zirconium and its alloys are good for thermal reactors. They have a low-capture cross-section for thermal neutrons and good thermal conductivity.

However, pure zirconia does not have good mechanical properties in E. Therefore, it is alloyed with vanadium and/or niobium. Zircaloy alloy used in Russian nuclear power is $Zr + (1-3)\%$ Nb. Doping seeks to improve not only mechanical but also corrosive properties.

Zirconium has great activity in hydrogen and O_2. When T is greater than 400°C, a vigorous reaction occurs with hydrogen: Zr dissolves hydrogen and zirconium hydrides are formed, in which the starting material is highly embrittled. With special technology, this propensity is reduced. Eliminating this disadvantage is completely is almost impossible. However, zircaloy is widely used in water-water energetic reactors (VVER) and reaktor bolshoy moshchnosty kanalny (RBMK) reactors.

3.3.4 Stainless steel

Steels has good mechanical properties but lower thermal conductivity than other materials and a high thermal neutron capture cross-section. It is corrosion-resistant in water at 360°C and in steam, gases, and LM to 650°C.

The diseased thermal neutron capture cross section is not of particular importance for fast neutron reactors. Therefore, stainless steel is the basic material of construction for such reactors; there is good experience with their use in thermal reactors.

3.3.5 Graphite

Graphite is used for the shells in the fuel vapor of high-temperature gas-cooled reactors. High thermal conductivity and a very low thermal neutron capture cross-section are great advantages of this material. Graphite is technological. From this, it can be pressed to obtain products of various geometric shapes and is easy to process mechanically. Graphite is well-compatible with fuel. However, it requires strict adherence to the gas mode, because graphite burns.

To ensure the tightness of graphite claddings, coatings of pyrolytic carbon and silicon carbide are used. Pyrocarbon is the form of graphite in which most atoms are arranged in the form of parallel layers. It retains gaseous fission products well, such as Xe, Cr, and silicon carbide, as a barrier effective for solid fission fragments.

3.4 Coolant

The primary coolant of the reactor cooling circuit directly cools the fuel rods.

There are basic requirements for material coolants:

1. You should have a small neutron capture cross-section in the operating range of the reactor.
2. You must have good thermal properties to ensure effective efficiency.
3. The energy cost of pumping must be small
4. There should not be corrosive and erosive activity.
5. There should be a little radiation-activated reactor.
6. It must be resistant to radiation and heat flow (i.e., should not decompose under their influence).
7. The safe operation of plants must be ensured (not explode, not toxic, etc.).

Light and heavy water (H_2O, D_2O), gases (CO_2, He, etc.), liquid metals (Na, Li, K, Pb, Pb-Bi and others) are used as heat carriers.

3.4.1 Water

Light and heavy water differ in nuclear properties. This is important when using them as moderators. The requirements for moderators will be discussed further. For the rest of their properties, these two materials are the same. Therefore, as a coolant, H_2O and D_2O are the same (except for their values).

Water has great heat capacity, so it requires acceptable energy for pumping. An important advantage of water over other coolants is that it is the only working fluid in the steam turbine cycle. Consequently, it is possible to create steam directly in the reactor and feed it to the turbine.

Disadvantages are that:

1. The water is corrosive and erosive. It is therefore important to maintain the water regime (water quality).
2. When a crack occurs, the uranium fuel rod interacts with water.
3. Water is (still) at the low boiling point. Therefore, it requires the creation of high pressures in the paths if you want to have good steam parameters at the turbine.
4. As a good retarder, water may not be useful for fast breeder reactors.
5. Under irradiation, it may be subjected to water radiolysis (decomposition of O and H).
6. Water is activated in the neutron flux.

We discuss the last two drawbacks in detail. During radiolysis, oxygen and hydrogen are released according to the scheme:

$$2N_2O \leftrightarrow 2N_2 + O_2.$$

As a result, if not avoided, this may form an explosive mixture. In addition, activated oxygen oxidizes in the core structural materials and may force them beyond the core.

Oxygen and hydrogen occur after nuclear reactions with the formation of radioactive isotopes:

$$^{16}O\,(n, p)\,^{16}N + \beta^-, \gamma\ (7.35\ s.)$$

$$^{17}O\,(n,p)\,{}^{17}N + \beta^{-}\,(4.14\ s.)$$

$$^{18}_{8}O\,(n,\ \gamma)\,{}^{19}O \rightarrow {}^{19}_{9}F + \beta^{-} + \gamma\,(\sim 29\ s.)$$

$$^{1}_{1}H + n \rightarrow {}^{2}_{1}H + n \rightarrow {}^{3}_{1}H\,(\sim 12\ year)$$

Provided that even a residual amount of impurities in the form of salts is present in water, such nuclear reactions take place in sodium and calcium:

$$^{23}Na\,(n,\ \gamma)\,{}^{24}Na + \beta -,\ \gamma\,(15\ h.)$$

$$^{44}Ca\,(n,\ \gamma)\,{}^{45}Ca + \beta -\,(164\ day)$$

Parentheses denote the half-life of an isotope.

Thus, in the coolant (water) of the first circuit, even they do not fall, fission fragments accumulate radioactive substances, including gases.

In addition to good water treatment, this requires additional devices: gas removers, chemical water purification devices, and so on.

3.4.2 Gases

Because, unlike water, compressible gases and boil, gas temperature limitation is not related to pressure. It is limited only to the gas resistance and the materials used. In the nuclear power industry, carbon dioxide (CO_2), hydrogen, and helium are mainly used. Advantages of gas as a coolant are that they:

1. are weakly activated reactor radiation and almost do no corrode (e.g., as a neutral gas, they are not generally noncorrosive and not activated);
2. do not slow (almost) neutrons (i.e., they can be used in fast breeder reactors);
3. make it possible to obtain high temperature (1000°C or higher) and have highly efficient energy.

Disadvantages of gas are that they:

1. They have low heat capacity and heat dissipation; therefore, to increase the gas density, it is necessary to have high pressure in the circuit;
2. They require a lot of energy to pump;
3. Coolants are in thermal reactors, which requires a separate retarder.

3.4.3 Liquid metal

In principle, coolants in nuclear power engineering may be the LM Na, K, Li, Na−K alloys iPb-Bi, or Hg.

They do not slow (almost) neutrons and can be used in any type of reactor. Thermophysical properties (thermal conductivity and heat) are good, but the heat capacity is worse than that of water, so they require a bit more energy for pumping. The main advantage is the possibility of obtaining large-thermodynamic cycle temperatures at substantially atmospheric pressure owing to the high boiling points.

3.4.4 Disadvantages

1. LM is strongly oxidized; therefore, systems with them require absolute tightness;
2. Na, K, and Li react violently with water;
3. LM requires complex systems treatment and heating when a reactor shuts down:
4. Eutectic Pb—Bi activating radiation reactors form polonium, which is highly toxic (mercury is toxic in general);
5. Inert gases are used as gaseous media in compensating tanks with a liquid metal coolant.

3.5 Moderator

The main purpose of moderators is to slow the neutrons to the required energies. Therefore, the two main qualities are:

1. High-retarding ability, which is quantitatively measured as $\xi\Sigma s$, where ξ is the logarithmic average loss of neutron energy at a collision with a nucleus retarder; and Σs is the macroscopic cross-section for thermal neutron scattering;
2. Small absorption neutrons, with a quantitatively measured moderator coefficient $\xi\Sigma s/\Sigma a$, Σa absorption cross-section of the moderator.

The retarder is better than both of these quantitative characteristics. However, nature is constructed so that an increase in one, as a rule, leads to a decrease in the other.

The neutron-physical data for the most commonly used inhibitors are provided in Table 1.3.

Light water has the highest retarding power, because, as the most light isotope, hydrogen has the best scattering properties (Table 1.3).

The ability to retard light water is about 10 times higher than severe. However, heavy water has the highest deceleration rate that exceeds the same performance of a conventional water ~ 50 times, owing to the extremely low neutron absorption cross-section.

The data in Table 1.3 also provide some conclusions about the design features of reactors using one or another moderator.

1. Because the deceleration rate of D_2O is the largest, the nuclear fuel is used more economically in heavy water reactors, because the unproductive losses of

Table 1.3 Neutron-physical characteristics of main inhibitors.

Material and characteristics	D_2O	H_2O	C	Be
$\xi\Sigma s$ (cm^{-1})	0.1280	1.160	0.0494	0.1561
$\xi\Sigma a$ (cm^{-1})	3.9×10^{-5}	1.9×10^{-2}	2.6×10^{-4}	8.2×10^{-4}
$\xi\Sigma s/\xi\Sigma a$	3300	61	190	~ 190
$1/\xi\Sigma s$ (cm)	7.8	0.86	20	6

neutrons in them is considerably less than in reactors of other types. This allows the use of natural uranium in reactors at sufficiently high specific loads. After that, C12 and Be9 can be regarded as economically viable moderators.

2. They have slowing abilities above all in ordinary water. This means that with the moderator, rods can make the bars more closely. Furthermore, according to the degree of tightness of the lattice, the order is Be, D_2O, and then C. Thus, the most compact at the same power can be a thermal reactor with H_2O as moderator.

Be is too expensive a material; it is stationary in nuclear power and is not used. It has found a use in space nuclear power as a reflector material.

Materials retarders should also possess such qualities as:

1) good mechanical and processing properties;
2) compatibility with other materials of the reactor;
3) resistance to thermal and radiation effects, and so on.

3.6 Absorber

The reactor uses the effect of changing the multiplication coefficient by introducing and removing neutron absorbing / multiplying materials to control the chain reaction. Devices implementing operative, targeted alteration multiplying properties of the reactor core are called regulators. In case of any emergency situation requiring immediate shutdown of the reactor, use special devices such as emergency curled bodies, which also use a multiplying effect of changing the properties of the reactor by injection (withdrawal) absorbers (breeder).

For the reactor to work for a sufficient time between overloads, more fuel is loaded into it than is required to maintain criticality. To compensate for the excess multiplication factor, special devices are used, based on the same principles as described for compensation bodies. All of these devices have a common name, organs of influence, on the neutron multiplication rate.

Stuckey, the apparatus containing materials that absorb neutrons, is an integral part of the reactor.

The basic designs of regulatory bodies, protection, and compensation do not differ much from each other. Moreover, one and the same body, depending on the situation, can often be used to perform any of the three functions. The absorbers can be used in the reactor in all three states: solid, liquid, or gaseous. Generally, for solid absorbers used in devices, a typical structure is schematically shown in Fig. 1.13.

It is understood that the 1th embodiment effect is obtained only by injection of the absorber. Placed in the core and prepared under the sink, nothing is taken. In the second embodiment, the effect is achieved by entering the retarder (scattering) absorber. Compared with the first option, neutron leakage is reduced with an incompletely introduced absorber, because the void is occupied by the moderator. In the atom version, neutron saving is the main goal. In the third embodiment, the impact

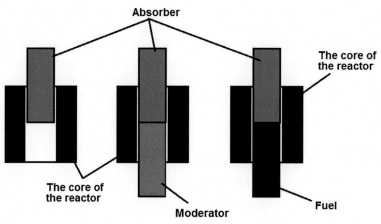

FIGURE 1.13

Typical design of rod reactor control system.

on the efficiency of the reactor multiplication factor markedly increases compared with the first two embodiments by the simultaneous action of two effects: the fuel substitution absorber reduces duplication and increases absorption in the active zone.

Design requirements bodies have an impact on the multiplication factor and absorbent materials.

The constant effectiveness of the organ affecting the reproduction coefficient must be ensured during its long-term operation in the reactor core. Absorber burns out, so this is accomplished when the agencies have enough fat to get absolute black body for neutrons. Changing the efficiency in this case can only be associated with a change in the geometric body sizes.

Ensure dimensional stability and thermal and mechanical properties. Under the action of neutron absorbing material, nuclear reactions occur, resulting in new elements, in which

a) the density and the geometrical dimensions of parts may vary, especially if the result is a gaseous element; hence, swelling of parts have to be considered in the design, allowing for necessary clearance;
b) thermophysical and mechanical properties of materials absorbers may vary;
c) corrosion resistance may vary material (absorbers).

Requirements for thermal characteristics (thermal conductivity) are important, because upon absorption, neutrons release as kinetic energy of particles b and y, which are eventually converted into heat, which must be removed efficiently. The requirements for the mechanical properties depend on whether the design of the materials absorber has a bearing. The requirements for corrosion resistance should be satisfied so as not to destroy the activity of part bodies and not to contaminate the coolant corrosion products, which have high activity.

Table 1.4 Properties of main absorbent materials.

Material	σ_a (barn), thermal neutrons	Resonance integral (barn)	Reaction	T melting, °C	Note
B^{10}	3840	–	(n, α)	2300	
B^{11}	755	280	(n, α)	2300	
Cd	2450	–	(n, γ)	321	
Hf	105	1800	(n, γ)	2220	
Gd	46,000	67	(n, γ)	1350	Rare earth element
Eu	4300	1000	(n, γ)	900	Rare earth element
Sm	5600	1800	(n, γ)	1052	Rare earth element

Isotopes of boron (B), hafnium (Hf), cadmium (Cd), gadolinium (Gd), europium (Eu), and samarium (Sm) are most strongly absorbed by neutrons. You can compare the absorbing and other properties of various absorbers using the data in Table 1.4.

3.6.1 Boron

Boron is most often used in nuclear power as an absorber material. Natural boron is composed of ∼20% B^{10} and ∼80% B^{11} with absorption cross-sections and 3840 σ end ∼0.05 σ, respectively (i.e., the absorber is B^{10}).

Natural boron is enriched by B^{10} before 90% to increase the efficiency of the absorber:

$$^{10}_{5}B + n \rightarrow {}^{7}_{3}Li + {}^{4}_{2}He + \gamma + \sim 3\,MeV\ kinetic\ energy$$

$$^{10}_{5}B + n \rightarrow {}^{3}_{1}H + 2\,{}^{4}_{2}He + \gamma + \sim 3\,MeV\ kinetic\ energy$$

Thus, the reaction of (n, α) to boron is in two channels. Moreover, the second channel core also produces tritium, which is a radioactive half-life of ∼12 years.

As a result of absorption by boron neutron α-large particles are obtained with large kinetic energy and a small path length:

1) absorbers made of boron or its compounds are heated and their organization require cooling;
2) α-particles with prolonged irradiation of the boron neutron disrupt the structure of the material, resulting in formable structures made of them.

Boron is used in the amorphous or crystalline form, usually consisting of the compounds borax ($Na_2B_4O_7$), boron carbide (B_4C), boron nitride (BN), and boric acid (H_3BO_3). Boric stainless steel containing ∼0.5−2.4% B is common. It has sufficient corrosion resistance during operation of the reactor, and satisfactory technological properties. However, appreciable forming parts of it do not allow the use of this steel for bearing structures. Because of this, it is necessary to install parts to provide for a possible increase in their size.

Boron carbide B_4C is a chemically stable compound with a melting point of 2450°C. Important properties are that it is not sublimated and does not burn at temperatures up to 2450°C. The main way to obtain parts is by pressing it from a powder followed by sintering. Typically, tablets are pressed from B_4C.

In reactors of pressurized water, which will be discussed later, liquid absorbents used (typically in the form of boric acid H_3BO_3) are mixed into the coolant. When using boric acid, as we have seen, radioactive tritium forms. It replaces water in the ordinary hydrogen atom. The result obtained is T_2O, HTO, and free hydrogen. Deleting associated tritium from the water is a serious problem.

3.6.2 Cadmium

Cadmium is a good absorber of thermal neutrons from the reaction:

$$^{113}_{48}\text{Cd} + \text{n} \rightarrow \text{n} \rightarrow {}^{114}_{48}\text{Cd} + \gamma$$

However, the low melting point and poor mechanical to give it a wide range of applications. Cadmium is mainly used in experimental work in research reactors.

3.6.3 Europium

This element is interesting because it is an effective absorber of neutrons in the thermal and thermal fields of energies. During neutron capture, its two natural isotopes, ^{151}Eu and ^{153}Eu, form a chain of transformation into other isotopes, which in turn absorb neutrons well.

3.6.4 Reaction

$$^{151}\text{Eu} + \text{n} \rightarrow {}^{152}\text{Eu} + \text{n} \rightarrow {}^{153}\text{Eu} ,$$

$$^{153}\text{Eu} + \text{n} \rightarrow {}^{154}\text{Eu} + \text{n} \rightarrow {}^{155}\text{Eu} + \text{n} \rightarrow {}^{156}\text{Eu} \, (\sigma_a \approx 0) .$$

Europium is used as Eu_2O_3 oxide with a melting point of 2000°C or greater. Europium is used to manufacture burnable poisons, because its efficiency is high and it can be made of thin rods or membranes.

3.7 Classification of nuclear reactors

The foregoing discussion provides a reasonable basis upon which to classify reactors.

1. Intended use

Designed for the production of secondary fuel (Pu^{239} or U^{233}).
Energy for thermal and/or electric energy (or mechanical work).
Dual purpose: to produce new fuel and to produce energy.

Research for powerful neutron fluxes, although thermal power does not exceed 50−60 MW and require special cooling of the organization.

Critical assembly for model studies of different configurations of reactors. Power is not more than a few kilowatts of cooling and special organizations are not required.

2. According to the neutron spectrum:

Heat, intermediate, or quick.

3. According to the moderator:

Graphite (RBMK, AM, AMB. HTGR, Wilf, etc.).
Light water (VVER. VK, PWR, BWR, etc.)
Heavy water (Canada deuterium uranium reactors).
Oxides or beryllium (special-purpose reactors).
AM - The world's first nuclear power reactor. It was commissioned in 1954 in Obninsk, Russia.
AMB - Water-graphite channel reactors at Beloyarsk NPP, Russia.
VTGR - High-temperature gas-cooled reactor.
VK - Experimental boiling water-water reactor.
PWR - Pressurized water reactor.
BWR - Boiling water-water reactor.

4. By way of placing the fuel in the moderator and/or whether the core is;

Homogeneous (gas phase or liquid).
Heterogeneous (RBMK, AM, or AMB).

5. By the nature of the coolant:

Gas, water, LM, or with organic coolant.

6. The principle of use:

Stationary, transportable, transportation (icebreakers, submarines, or NRE).

7. By design:

Vessel: inside the case, a common flow of coolant flows. The vessel is loaded by internal pressure coolant.

Channels: coolant flows through each channel separately with the fuel assembly; the body is not loaded with pressure fluid. This load is technological channels.

Pool: vessel, a large tank, unpressurized, in which at a sufficient depth under water (several meters), the reactor core is located, through which this water is pumped.

The design of thermal reactors is determined by the pair "coolant-moderator":

1) water-cooled-only vessel type;
2) gas-graphite-only vessel type;
3) water-graphite-only channel type.

4. Conclusion

This chapter presented the fundamental issues arising from the design of nuclear reactors and of their appearance. Special attention is given on the construction of two main elements of active zones of all reactors fuel and bodies, and their impact on the multiplication factor. The chapter discussed the basic requirements necessary for these materials of construction, as well as the most important properties of these materials. The classification of nuclear reactors was introduced in the presentation of terminology well-established in the nuclear power industry. The knowledge acquired is enough to proceed directly to the study of the logic of justification designs of nuclear reactors.

Suggested reading

Akkerman, G., Adam, E., Kabanov, L.P., 1984. In: Margulovoy, T.H. (Ed.), Efficiency and Safety of Nuclear Power Plants (VVER): Textbook. M: Higher School.

Budova, V.N., 1985. VA Ferafonov Construction of the Main Equipment NPP: A Manual for Schools. M: Energoatomisdat.

Dementiev, B.A., 1984. Nuclear Power Reactors: A Textbook for High Schools. M: Energoatomisdat.

Emelyanov, I.Y., Mihan, V.I., Solonin, V.I., 1982. Under the General. In: Dollezhal, A.O. (Ed.), Engineering Nuclear Reactors: Training Benefit for Schools. M: Energoatomisdat.

Novikov, V.M., Slesarev, I.S., Alekseev, P.N., 1993. Nuclear Reactors Increased Security (Analysis of Conceptual Designs). M: Energoatomisdat.

Ran, F., et al, 1989. In: Legasova, V.A. (Ed.), Handbook of Nuclear Energy Technology. Translated from English. M: Energoatomisdat.

Styrikovich, M.A., Shpilrain, E.E., 1981. Energy. Problems and Prospects. M: Energiya.

Walter, A., 1986. A Reynolds Breeder Reactor on Fast Neutrons. M: Energoatomisdat.

Non-power applications—new missions for nuclear energy to be delivered safely and securely

William J. Nuttall, PhD, BSc, Gareth B. Neighbour, PhD, BSc

School of Engineering and Innovation, The Open University, Milton Keynes, United Kingdom

1. Introduction

This book looks to the future of nuclear technology deployed for civil and commercial purposes. The preceding chapter has introduced the technical fundamentals that have shaped today's nuclear industry including matters of energy conversion efficiency and materials engineering. In this chapter, we explore the future of nuclear technology away from its historical orientation on the base-load generation of low-carbon electricity. We introduce some potentially attractive commercial opportunities outside electricity and point to the possibility that "buyers" may be found with a greater willingness to pay and a more open approach to matters of sociotechnical and economic risk. Within this approach, we pay particular reference to safeguards and security. Key questions are proposed with the recognition that subsequent chapters will focus on specific issues relating to nuclear research and nuclear medicine, so those important aspects are not explored in detail in this chapter.

2. New opportunities for civil nuclear energy

2.1 Historical background

From its earliest days, the technology related to civil nuclear energy has provided more than simply electricity. Indeed, the small (5 MW_e) facility located in Obninsk 110 km southwest of Moscow was the reactor with arguably the strongest claim to be the world's first nuclear power plant. For most of its life, a facility for the production of radioisotopes and for nuclear research, rather than a source of useful electricity. The Obninsk Atomic Power Station 1 at the State Scientific Center AJ Leipunsky Institute of Physics and Power Engineering was first connected to the grid in June 1954, but the electrical connection was cut just 5 years later. However, the reactor continued to operate, devoted to its nonelectrical role, until 2002.

Rachkov et al. (2014) presents an account on how this transformation occurred and how it was instrumental in later generation of nuclear power moving from military to peaceful purposes.

Sweden was an early adopter of nuclear power and one of the first to seek to utilize nuclear energy for district heating purposes (Csik and Kupitz, 1997). The Ågesta nuclear power plant near Stockholm started commercial operation in 1964. It operated for 10 years providing 68 MW_{th} of district heating to the Stockholm suburb of Farsta together with a modest amount of electricity (12 MW_e) for the Swedish grid. While the reactor's role in nonelectrical heat provision is historically important, especially for the development of industrial process heat, it must be conceded that the Ågesta reactor's primary purpose was to convert its natural uranium fuel into plutonium for the short-lived Swedish nuclear weapons program. This aspect is discussed by Paul (2000) in *"Power versus Prudence: why nations forgo nuclear weapons"* in which he argues for a prudential-realist model where a nation's national nuclear choices depend on specific regional security contexts. However, Jonter (2010) presents a compelling case that while Sweden would have been capable of producing nuclear weapons within a few years of 1968, at the point Sweden signed up to the Nuclear Non-Proliferation Treaty (NPT), there was a set of complex reasons that saw Sweden withdraw from its military aspirations restricting future activity to solely civilian aspects, including the focus on district heating. The reasons motivating the Swedish decision included rising public opposition to nuclear weapons, tension between civilian nuclear power goals, and the desire to maintain "freedom of action," the pressure from United States to discourage proliferation and the strengthening of international nonproliferation norms (e.g., the NPT opened for signature in 1968 and came into force in 1970). Returning to the theme of district heating, other countries including the former USSR and Romania followed Sweden in deploying cogeneration systems powered by a nuclear reactor. Perhaps, the most interesting example is the small Bilibino nuclear power plant in northern Siberia (Horak, 1997). This station comprising four 12 MW_e EGP-6 reactors has provided power for a seasonally isolated gold mining town since 1974. At its peak, the Bilibino plant produced 78 MW_{th} of useable district heat. At the time of writing in 2020, three reactors continue to operate, but these are expected to close shortly. It is planned that the role of the Bilibino plant will be replaced by the new Russian floating nuclear power plant, the *Akademik Lomonosov* (Zverev et al., 2019). In Romania, the Cernavoda heavy waterpower reactors (examples of Canadian CANDU-6 technology) provide district heating for the town of Cernavoda. When in use, the heat needs of the small town represent a negligible load on the nuclear power plant, which still primarily sheds its low-grade heat to the Danube-Black Sea Canal. Cholewinski and Tomkow (2017) comment on the extent of cogeneration and the importance of external district heating in various parts of Europe including Cernavoda and suggest that nuclear power plants can be a reliable source of heat supplying district heating systems. Importantly, they argue that the thermal characteristics of pressurized water reactor (PWR) technologies are sufficient to supply a district heating network and cover the situation for future cogeneration systems in, for example, Poland. They add that PWR technologies are especially viable as replacements for traditional coal-fired cogeneration units and hence could help in the fight against greenhouse gas emissions.

Another important nonelectrical application of nuclear energy has been seawater desalination. General principles of the technology and the use of nuclear energy can be found in Khan et al. (2018). Reflecting on the history, Guth (2018) describes how the city of Shevchenko/Aktau near the Caspian Sea provides an example of Soviet technopolitics from 1950s, including roles ranging from the mining of uranium for the Soviet A-bomb to a testing ground for fast breeder reactor and ultimately to a nuclear desalination site present day. It is another example on how military intentions can result in civilian benefits over time, consistent with the classical innovation "S" technology curve. Prominent in that role is the interesting case of the BN-350 sodium-cooled fast reactor at Aktau, formerly Schevschenko, in Kazakhstan (Nuttall and Storey, 2014). Desalination can be achieved at scale by diverting only a small proportion (less than 10%) of the nuclear steam. The Russian nuclear industry leads the world in deploying such technology. Most recently, an installation capable of producing $10{,}200 \text{ m}^3$ of freshwater per day was commissioned at the Russian-designed Kudankulam nuclear power plant in India, as reported by Rosatom (2020).

The largest nonelectrical role for civil nuclear technology has been related to the use of small reactors, which might loosely be termed "research reactors" (Nuttall and Storey, 2014). The World Nuclear Association reports that approximately 220 research reactors are currently operating in 53 countries (WNA, 2020). The term "research reactor" typically covers a range of nonpower applications:

- Accelerated materials testing (e.g., nuclear fuels and structural materials—for nuclear and nonnuclear applications).
- Training (e.g., for power plant operators, nuclear regulators, and university students in nuclear engineering and cognate disciplines).
- Isotope production (e.g., for medical applications in diagnosis and therapy and for industrial purposes).
- Scientific research (e.g., the use of neutron beams in condensed matter science).

Typically, research reactors do not generate electrical power. Historically, many have made use of highly enriched uranium (HEU), i.e., with fuel containing 20% or higher concentration of ^{235}U (see later chapters), either as reactor fuel or as "targets" for medical isotope production.

Looking internationally, we note the development of new medical and industrial isotope facilities such as the new ANSTO Nuclear Medicine (ANM) facility in Australia. Isotopes such as ^{99}Mo, which is the parent isotope of ^{99m}Tc, are made by the irradiation of uranium targets. ANSTO estimates that ^{99m}Tc is used in 85% of all nuclear medicine procedures in Australia and the most common in nuclear medicine worldwide (WNN, 2020). The new facility in Australia has the potential to create significant exports globally, which brings an interesting dimension regarding the limit to 20% ^{235}U in low enriched uranium (LEU). The limit was fixed at 20% and is increasingly appearing to be somewhat artificial (Glaser, 2005; Ashley et al., 2012;

Brown and Glaser, 2016). Fundamentally, medical isotopes are produced in "research" reactors and processed in special facilities such as ANSTO's Lucas Heights site near Sydney (which in 2012, the Australian government invested AUS168.8 million in the construction of the nuclear medicine manufacturing plant associated with the OPAL research reactor).

All of the above are of course set in the context of nuclear generation. To illustrate the scale, in the case of the European Union (2019), there were 128 nuclear power reactors (with a combined capacity of 119 GW_e) operating in 14 of the 28 Member States accounting for over one-quarter of the electricity generated in the whole of the European Union (FISA, 2019). Nuclear currently accounts for 53% of the EU's carbon-free electricity. More globally, there are around 445 nuclear power reactors in operation in 30 countries, which should be appreciated for generating immense amounts of clean power. It is important to recognize that the current level of nuclear energy supply avoids more than 2.5 billion tons of carbon dioxide emissions annually had the same amount of electricity been generated using fossil fuels. It is evident how potential industrial uses of nuclear energy are closely coupled with those states pursuing a civil nuclear agenda and the advantages this brings in a wide range of industries (IAEA, 2016a).

2.2 Looking ahead

The posited nuclear renaissance of the early 2000s has struggled to occur in western liberalized economies. In part, this may be attributed to the global financial crisis of 2008/2009 or to the severe nuclear accident at the Fukushima-Daiichi plant in Japan that arose as a consequence of a major earthquake and tsunami in March 2011. Both of these were on the back of the uncertainties associated with final disposal of nuclear waste, which created anxiety in the general public and continued to struggle to find a resolution across the globe. Whatever the background challenges, the proximate concern for putative developers of new nuclear power plants has become that the costs of new build (physical construction and financing) are simply too high and far above the market's willingness to pay. Policy in the United Kingdom has sought to increase the amount that will be paid to investors and to remove market price volatility from investors, via a rebuilding of the wholesale electricity market and especially through the introduction of long-term "contracts for difference" (CFD) for low-carbon generation. However, even this measure is proving insufficient to motivate developers with the CFD prices that are believed to be on offer. First Toshiba in 2018 and then Hitachi in 2019 pulled back from plans to develop new nuclear power plants in the United Kingdom once it became clear that the level of CFD support that had been won by EDF Energy for its consortium developing the Hinkley Point C nuclear power plant would not be made available for later projects. The consequence of these realities is a widespread recognition that as revenues in electricity cannot be as high as once foreseen, the costs of new build must fall. Indeed, the UK Industrial Strategy's Nuclear Sector Deal of 2018 states (Nuclear Industry Council, 2018):

The Nuclear Industry remains committed to a journey to reduce costs across the sector and are setting out a target within this sector deal of reducing the cost to the consumer of low cost low carbon electricity from nuclear new build, based upon the actions in the sector deal being delivered by government and industry together, by up to 30% up to 2030.

It has become orthodox thinking among those concerned for nuclear power new build that capital costs must be reduced to restore economic competitiveness, especially noting the falling cost of competing low-carbon renewable sources of electricity and the technologies necessary to improve the dependability of renewables-based power.

While we do not suggest that UK Energy Policy has been without fault or difficulty, there have been some attributes that we suggest have been beneficial and could be of interest to other territories. For example, within the European Union, the United Kingdom was an early, and relatively consistent, advocate of policy goals accompanied by technological neutrality. That is, a policy goal might be for low-carbon generation, but the means to best achieve that should, as an aspiration at least, be left to investors. In contrast, many EU countries and the EU itself tended to favor technological prescriptions such as for a minimum contribution from renewable energy to the exclusion of other forms of low-carbon power generation, such as nuclear energy. As an EU member state, until BREXIT at least, the United Kingdom would find itself forced into technologically determined solutions by EU requirements irrespective of a more local preference for technological neutrality. Within EU policy-making, the United Kingdom maintained a strong voice for best science in policy-making and for market-determined solutions, consistent with public policy goals. Indeed, which type of nuclear power plant a power company might wish to build would be a matter for the investors to decide.

In safety regulation, the United Kingdom also takes a rather liberal and nonprescriptive approach. We shall say more about the UK approach to safety later, but one important aspect is a consequence of a major piece of public policy in the early 1970s. The UK Committee on Health and Safety at Work was initiated by Barbara Castle, Employment Secretary in Harold Wilson's Labour government. The committee was led by Lord (Alfred) Robens, Chairman of the National Coal Board and a former trade unionist. The Robens Report, as it became known, was published in June 1972, and it heavily influenced the UK Health and Safety at Work Act of 1974. That legislation paved the way for the notion that safety is the responsibility of the company concerned, as the operator. Furthermore, it is appropriate, and indeed desirable, if safety regulators merely set the standards to be met and assess the viability of proposals from operators rather than actually dictating the actions that should be taken. In this way, those closest to business operations (the operators) preserve and maintain ownership and direct implementation of safety. Most other countries, including the United States, have never taken such a step, and safety-related actions are typically mandated in specific detail by regulators. Lord Robens' ideas were adopted wholeheartedly by the UK Nuclear Installations Inspectorate, and UK nuclear licensing still pursues a relatively nonprescriptive approach in the United Kingdom.

In principle, one might expect liberal markets and nonprescriptive policies and regulations to favor private investment in nuclear energy technology. That was arguably the hope lying behind UK talk of a Nuclear Renaissance in the 1990s. As things turned out, however, the EU-ETS carbon price that had been anticipated to inhibit enthusiasm for natural gas—based power generation failed to rise, as expected. Nuclear technology costs were seen to increase in the first decade of the 21st century, and potential investors had a strong sense that economic risks could not be adequately hedged. Finally, the banking crisis of 2008—09 reduced the availability of capital, and the consequent recession led to a significant drop in electricity demand. This loss of demand eroded confidence for the building of any power generation assets, especially costly and risky options such as nuclear power plants. In consequence, by the time of the Fukushima-Daiichi nuclear accident in March 2011, the Nuclear Renaissance was already in a perilous state in liberalized markets such as the United Kingdom and parts of the United States. After the loss of public trust seen in much of the world (but interestingly not so much in the United Kingdom) after the Fukushima accident, nuclear new build policies everywhere were in seen to be in trouble. Nuclear energy as a means to generate low-carbon electricity increasingly appeared less attractive to investors and politicians than renewable alternatives, notwithstanding the problems (e.g., reliability) associated with such alternative technologies. Consequently, one may also argue by the coupling of nonpower applications to generation; this aspect also took a "hit" in progressing Generation IV reactor technology and the possible impact on the hydrogen economy, for example.

Given the difficulties faced by nuclear new build in the United Kingdom oriented to base-load electricity production, we suggest that alternative ideas are required. Charles Forsberg at the Massachusetts Institute of Technology is one voice speaking of another way ahead for nuclear energy. Rather than focusing solely on reducing costs of electricity generation, the nuclear industry should pay more attention to increasing its revenues. He has particularly focused on the possible benefits of coupling high-temperature nuclear reactor technology to high-temperature heat storage so as to facilitate flexible electricity output better suited for meeting price varying demand (Forsberg et al., 2018). One of us (WJN) is also associated with similar research oriented to lower-temperature light water reactors (Wilson et al., 2020).

Japan has sought to take advantage of novel business models for nuclear energy, especially in regard to the hydrogen economy, through the development of the high-temperature test reactor (HTTR) but has failed to deliver the promised yields; indeed, there has been a long history of graphite-moderated high-temperature reactors that have promised so much in terms of cogeneration (Kugeler and Zhang, 2019). There is also the interesting dynamic of modularization of the reactor technology. This can bring down costs in manufacturing; make it easier to place generation nearer to the demand; and potentially link to carbon sequestration as well as hydrogen production (IEA, 2017).

Building upon a desire to enhance revenues takes us to consider who might be willing to pay more for the services that nuclear technologies can provide. A corollary to such thinking is the observation that increasingly: *low-carbon electricity is simply too easy and too cheap a task to motivate the use of nuclear technology.* That is, we suggest that the nuclear sector needs to deploy itself so as to tackle more difficult and important challenges. Typically, the resulting opportunities will be in domains where there is a greater willingness to pay. In this spirit, we suggest that one important place to look for new opportunities is the emerging policy goal of "deep decarbonization." In addition, we are led to suggest that nuclear technology should be focused on achieving the goal of decarbonizing the last 20% of whole energy system and not continue to be wedded merely to addressing the first 30% of the problem (low carbon electricity).

Many UK energy policy-makers have been attracted to the notion of a major electrification of industry, domestic heating, and mobility as being central to the path to a low-carbon future. For example, the push toward electric vehicles (over hybrids) is a brave one considering the lack of infrastructure and the need for critical minerals to support the battery markets. If such views prevail, it is possible that the major impact of UK nuclear energy technology will be in power generation, but a massive expansion of electrification is not inevitable. For example, there is currently much interest in expanded heat networks (with a particular project underway in Stoke on Trent) (Vital Energi, 2020). One of us WJN has, with AD Bakenne, pointed to another way ahead for a material step toward decarbonization by the United Kingdom: the development of hydrogen energy networks building upon today's natural gas industry and utilizing carbon capture, utilization, and storage (Nuttall and Bakenne, 2020).

The prospect of new energy supply markets (heat, hydrogen, ammonia, etc.) raises the prospect that new applications of nuclear energy technology might be identified. If opportunities can be found for sectors with few alternative options and with a significant willingness to pay, then it might be possible to establish a revenue source capable of handling the high costs and high economic risks associated with nuclear innovation. In that spirit, possible "hard"-to-decarbonize sectors with a willingness to pay that might exceed that of the low-carbon electricity sector include the following:

- Iron and steel
- Cement manufacture
- Marine propulsion
- Air travel

These options are being driven by the deep decarbonization agenda. That agenda has been advanced significantly in the United Kingdom by a major 2019 report from the Energy Transitions Commission (2020).

In particular, nuclear energy associated with low-carbon hydrogen production could help greatly with established chemical engineering techniques such as steam methane reforming and autothermal reforming to produce hydrogen from natural gas. To make a material difference to deep decarbonization, the nuclear sector should focus on high-temperature reactors engaged on process heat applications including hydrogen production, such as the HTTR example given earlier. Hydrogen production needs not be associated with water electrolysis or high-temperature thermochemistry but could be associated with today's existing fossil fuel and industrial gases industries providing that actions taken are all consistent with deep greenhouse gas emissions reductions (Nuttall and Bakenne, 2020).

Such process heat aspirations are well matched to the concept known as the very-high-temperature reactor (VHTR) proposed by the Generation IV International Forum (GIF) (2020). The GIF describes itself as "a cooperative international endeavor." The Forum originally convened at the initiative of the United States during the Presidency of George W. Bush currently (at March 2020) has 14 international partners working together to advance six fourth-generation nuclear energy concepts, one of which is the VHTR. One of the main drivers in favor of the VHTR concept has been the higher thermodynamic process efficiencies achievable at higher temperatures coupled with the material properties offered by graphite. However, the original 1000°C ambitions of the GIF for the VHTR concept have now been scaled back to 750°C recognizing the structural materials challenges at very high temperatures. The possible high efficiencies in electricity production at very high temperatures were indeed a motivation behind the UK's Second-Generation nuclear power program focused on the advanced gas-cooled reactor (AGR), which has a peak temperature around 650°C. With a gas primary circuit temperature above 600°C, this may indeed be regarded as a high-temperature reactor, but it is important to recognize that the moderator core is closer to 550°C, so limiting the thermal oxidation by the CO_2 coolant (but of course radiolytic oxidation remains).

The fact the United Kingdom commercialized such a technology 40+ years ago might suggest the country is in a good position for a leadership role in VHTR development, but the market became open and internationalized. This led to uncertainty on the preferred reactor technology for future electricity generation. Nevertheless, the VHTR is unique among the 6 GIF concepts under development in that it is focused on the potential for process heat applications as well as tried and tested graphite moderator technology. The GIF is also focused on the nongeneration benefits as depicted in Fig. 2.1. With that said, we must acknowledge that the breakup of, or change in, large centers of UK knowledge and experience such as UKAEA, BNFL, and NNC created a fragmented industry, which has in part been acquired by a sequence of overseas owners and seen numerous restructurings. Arguably, to an extent, the United Kingdom has been willing to buy in technology rather that develop its own. This choice of "buy" over "make" might be regarded as the proper operation of a market, but it can mask a wider failure of political economy—the need to sustain national competences. Lastly, the leadership potential for the UK nuclear sector from the past few decades has either moved into other industries or has retired.

The United Kingdom has in 2019 reinstated its funded membership of the US-led international GIF. The United Kingdom has a particular interest in the plans for sodium-cooled fast reactors, given past experience, and also for the very-high-

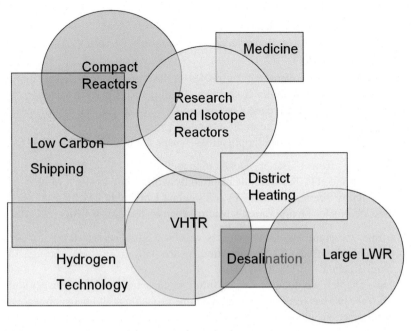

FIGURE 2.1 Generation IV nonelectrical opportunities.

Thermal nuclear technologies (circles) and technology application areas (rectangles), authors' own assessment. Electricity-based approaches are excluded, e.g., LWR power production for hydrogen via electrolysis. We note the potential future emergence of an overlap between compact reactor systems and VHTR. These are shown earlier to be separate. *LWR*, light water reactor; *VHTR*, very-high-temperature reactor.

temperature reactor, given the ideas presented earlier. Another aspect of deep decarbonization, beyond high-temperature process applications, would involve the use of small modular (light water) reactors, such as for civil marine propulsion purposes (see, for example, Peakman et al., 2019). One particularly exciting opportunity would be high-speed maritime services (as favored by nuclear propulsion). This would have the potential to take business away from air-freight—another hard to decarbonize sector; although it must be conceded that hydrogen also has the potential to transform aviation into a low-carbon mobility option.

In the area of civil nuclear technology, the UK government is advised by the Nuclear Industry Research Advisory Board (NIRAB). In its 2018/2019 annual report, NIRAB (2019) observes:

"Advanced Nuclear Technologies (Small Modular Reactors [SMRs] and Advanced Modular Reactors [AMRs]) could, in addition, maximise cost-competitiveness by satisfying a range of other needs within a wider decarbonised clean energy system, including:

- *Supply of low grade heat for domestic heating*

- *Supply of high temperature process heat to energy intensive industries*
- *Providing a source of energy to manufacture hydrogen*
- *Electricity supply to accommodate the intermittency of electricity generated from renewable sources"*

In 2018, the UK government published outline propositions from eight companies in receipt of feasibility assessment funding from the UK government for advanced modular reactor concepts (excluding light water reactor technologies) (Mathers, 2019). Three of the eight candidate projects are focused on the goal of high-temperature process heat applications. These are as follows:

- "U-Battery" from U-Battery Developments Ltd.
- DBD Limited HTGR ambitions inspired by work at the Institute of Nuclear and New Energy Technology (INET) at Tsinghua University in China
- Ultrasafe Nuclear Corporation's Micro-Modular Reactor (MMR)

In time, the eight feasibility projects will be whittled down to one, or perhaps two, technologies that will be supported by the UK government through to full commercialization.

In this chapter, we do not focus on future nuclear fusion—based energy sources. One of us (WJN) has long argued that the most probable path for fusion technology commercialization would lie not with electricity generation, but rather in process heat applications serving chemical engineering—based industry. One such example would be cryogenic liquid hydrogen production as a commercial product, but also potentially as a cryogen for the high-field superconducting magnets associated with future fusion "tokamak" concepts. Such a scenario was termed "Fusion Island" by Nuttall et al. (2005). Further detail was given by Nuttall and Glowacki (2008).

3. Security and safeguards and safety—the three S's
3.1 Security

In this chapter, we draw a distinction between matters of nuclear weapons proliferation and matters of nuclear terrorism. We regard the former as primarily relating to the activities of nation states under international safeguards and other arrangements to verify compliance with international obligations not to use nuclear materials for weapons (under the Treaty on the Non-Proliferation of Nuclear Weapons [NPT]). We associate the latter issue of terrorism and nuclear security primarily with non-state actors, although they may be state supported. Typically, the issues are less technically sophisticated, and the threat is perceived to be less substantial. It is sometimes said that the nuclear terrorist threat is more in the realm of weapons of mass disruption than weapons of mass destruction.

The canonical example of a nuclear security threat is the radiological dispersion device or "dirty bomb." Perhaps the other most widely considered scenario is an assault on a fixed nuclear asset, such as a power station. Such threats can be external (such as a truck bomb), internal (such as from a radicalized "lone-wolf" employee), or in-principle cyber in nature (noting the Stuxnet computer virus attack on Iran in 2010). To a great extent, and taking the UK's ONR (2014) as an example, these risks are considered in design codes that meet the safety assessment principles (SAPs). These SAPs take an objective risk approach as opposed to a zero-risk criterion. Other nation states have very similar principles and systems and may relate to over-arching 82 requirements (or principles) set out by IAEA (2016b). At the heart of the approach is to ensure the risk is "as low as reasonably practicable" (ALARP) or achievable (ALARA) in the United States. Nevertheless, such examples are, we suggest, sufficient to reveal many of the most pertinent policy issues and how nonpower applications can grow. We take the view that it is not in the public interest to speculate as to whatever other significant nuclear security threats might exist—they do! These two types of example show that the policy response to the risk of a nuclear security event is governed by regulatory systems in place coupled with national law enforcement and domestic security intelligence more than is the case with the threat of nuclear nonproliferation.

Nuclear security threats are by no means restricted to nuclear energy—related infrastructures such as power stations, fuel cycle facilities, or research reactors. For example, potentially very harmful radioactive sources are very widely used in hospitals (for example, in nuclear medicine for therapy and diagnosis and also for medical instrument sterilization) and by industry (such as in agriculture and food processing, metals processing, in mining and in the oil industry) as discussed earlier. The IAEA produces technical reports for such industries, and an example for food is report 482—Manual of Good Practice in Food Irradiation (IAEA, 2015). The extent of nuclear technology globally is significant and more divergent than ever before in the types of facilities and uses. In the rest of this chapter, we are drawn to the question: does the enhanced use of nuclear technology for nonpower applications give rise to new or more worrying nuclear security risks? While the notion of hazards and risk is well understood, perhaps what is less understood is how subjective or perceived risk influenced societal decisions over what objective analysis might suggest we do (see McGuire et al., 2010). It is probably true to say that nuclear security is less defined than the more societal concern on nuclear proliferation risks.

3.2 Safeguards—nuclear nonproliferation risks

In this section, we assess the risks posed by the proliferation of nuclear weapons and the threat of the use of nuclear weapons by sovereign states. One risk associated with nuclear proliferation is the notion that an aggressively minded state might deliberately seek to position itself such that it might relatively quickly "breakout" to nuclear weapons capability. One issue of concern in recent years has been whether a state might achieve breakout capacity without having infringed the Nuclear NPT, beyond simply its prohibitions of intent. Any state attracted to breakout capacity is likely to be located in a geopolitically difficult region of the world (in synergy to the arguments set out by Paul, 2000). Such geopolitical stresses help ensure that any clear move toward breakout capacity can propagate fear within the region and prompt arms races and dangerous perceptions of insecurity.

At the time of writing, much of the World's attention is focused on Iran in terms of safeguards. A contemporary example is the situation with Iran in which, since January 2016, the IAEA has been verifying and monitoring Iran's implementation of its nuclear-related commitments under the Joint Comprehensive Plan of Action (JCPOA) agreed in 2015 between Iran and the P5+1 (China, France, Germany, Russia, the United Kingdom, and the United States). Global politics is a strong influence, and since the United States has withdrawn and other countries may be forced to do the same at the time of writing, the heart of the issue relates to the LEU limit, but there are subtleties that perhaps are not obvious as outlined by Ashley et al. (2012) who interestingly ask the question "Is 20% LEU (although not weaponizable) a sensible upper limit for uranium enrichment assuming particular diversion routes?" The key in moving forward with safeguards is utilizing new technologies so that organizations such as the IAEA can continue to verify the nondiversion of nuclear material and evaluate, determine, and locate any absence of undeclared material and activity in breach of the NPT.

The history of the 20% limit is less known than might have been expected and seems, at least in part, to have been arbitrarily set at 20% to allow research reactors to do their work for nonpower applications. Brown and Glaser (2016) set out the key arguments from an historical perspective of how this limit originated and the significance of the limit. In essence, it was a political judgment to allow the operation of the research reactors and for them to be sufficiently productive rather than any logic associated with safeguards.

Key to prevention of nuclear proliferation, and associated risks such as breakout, is a robust Nuclear NPT and its additional protocols (enshrined in nation state's statutes). Within these arrangements are the requirements for "safeguards" and for intrusive rights of inspection, under the additional protocol. If materials detected with a higher value, does this indicate a higher risk? Arguably both nuclear technology and the approach to safeguards and security have changed little since the first days of the industry, but there are indicators that suggest a rethinking is needed, if some of the underlying issues and problems are going to be solved for future generations and especially if nongeneration applications and industries are going to grow to new economic levels. Part of that rethinking has involved a more robust inspections regime (the "additional protocol"), but what is being inspected remains remarkably unchanged. In terms of the Iran question, Iran has begun installing more advanced centrifuges and is moving toward producing enriched uranium, which might be considered illegal, but does this "forbidden" territory under its nuclear deal with major powers constitute a greater objective risk? The perception is clearly yes, but that is not the same thing as an actual risk. Of course, there have been significant advances in civil nuclear technologies, but the first questions are as follows:

- Has the evolution of technology reduced costs and risks?
- Should society, as a whole, feel more at ease concerning the risk of a severe accident, nuclear security attack, or proliferation event?

There is not space to cover the full range of technical measures, international conventions, or treaties that aim to protect against such occurrences here. A mere introduction and overview can only be provided, but in this chapter, we have hoped to present reflections that will help place in context much of what follows in the rest of this book.

3.3 Safety—severe nuclear accidents

No one can be under any doubt that the risk of a major event is nonzero. There are various approaches and models worldwide at estimating the risks, and in general, it is fair to say that the nuclear industry, even with the accidents that have occurred, has fared better than many. However, in the public's mind, it is at the forefront of thinking because of the nature of the risk. The idea of subjective risk was presented earlier, but just to place this in context, few in the western hemisphere remember the Union Carbide plant accident at Bhopal in 1984 killing in estimate 8000 people within a 2-week period of the release and affecting over 500,000 people exposed to methyl isocyanate. No nuclear accident has matched the scale of that disaster, whereas the IAEA has spoken of the expectation of 4000 accelerated deaths as a result of the world's worst nuclear accident at Chernobyl (IAEA, 2006, 2008). This tragic outcome, however, falls far short of the catastrophe represented by 8000 prompt fatalities.

Nuclear plants may have rigorous design codes, but the risk of a major accident is not infinitesimally small, say 1 in 10^6 per year, denoted as a basic safety objective, as stipulated in SAPs discussed earlier. In terms of nuclear accidents, there are four that come to the fore: Windscale (1957); Three Mile Island (1979); Chernobyl (1986); and Fukushima-Daiichi (2011). Chernobyl is by far the most significant. The European Parliament produced a 2016 briefing note in which it noted that 28 people died as a direct result of the accident in the first 3 months and another 19 died between 1987 and 2004 of various causes that were involved in the cleanup, but not necessarily associated with radiation exposure (EU, 2016). The European Parliament's briefing notes:

> *According to official estimates a total of 4000 eventual deaths from radiation-induced cancer and leukaemia can be expected among the higher-exposed populations, that is, the emergency workers from 1986 to 1987, evacuees and local residents of the most contaminated areas (this includes the workers who died of acute radiation syndrome and the children who died of thyroid cancer)*

The long-term health benefits and estimates of early deaths that might result remains hotly debated especially regarding the estimates of mortality outside Belarus, Russia, and Ukraine. Some estimates of all cancers and deaths are very much higher than the official estimates. Estimates are very dependent upon the methodology, e.g., linear, sublinear, superlinear, or hormetic dose-response relationships and whether the "one-hit" hypothesis is assumed. Such considerations must sadly remain beyond the scope of this chapter.

It is important to recognize that the SAPs and similar approaches are risk based and will assume a finite probability of accidents and consequent fatalities. Thus, the safety systems and mitigation strategies (or defense in depth) is significant to reduce the risk of a major immediate accident and typically 1 in 10^6 is used as a benchmark. On the other hand, the longer-term risks relate largely to pathways for radiological release, which are much harder to evaluate. This difficulty underpins much of the debate regarding the long-term consequences of Chernobyl. In particular, it prompts the argument over "concentrate and control" or "dilute and disperse," noting that they present different risk scenarios. The western paradigm based on humanist thinking is that "one death is too many" in the public's eye, i.e., the risk of one death that might result over 10^4 years (related to the UK's basic safety limit), coupled with key concepts such as the precautionary principle, and ALARP means that the "concentrate and control" is firmly the regulatory and legal stance. This can have perverse consequences if care is not taken. For example, if we apply this to graphite moderators worldwide, well in excess of 250,000 tons (IAEA, 2010), this means the need to create very large facilities in terms of volume for disposal (effectively storage) for 10,000 years with the required infrastructure and risk to terrorist activity, later generational "mining" and recovery of minerals, animal intrusion, etc. On the other hand, a mixed mode approach with controlled release of ^{14}C or even capture using modern day CCS technology would reduce the volume to less than 1% to a mineral ash, and any ^{14}C release would in effect be "neutralized" by the global carbon cycle in an effective half-life of 37.5 years (see White et al., 1985). Equally, a useful comparison can be made to the study by Hodgins (2009) measuring atomic bomb—derived ^{14}C levels in human remains to determine year of birth/or death where the impact was seen as small in predicting values. Another perhaps more famous case is in the forgery of vintage wine (ACS, 2010).

Society has, in general, failed to generate and realize an effective nuclear waste strategy partly because of the subjective nature of the thinking toward waste by a significant proportion of the public, based on the perceptions of nuclear safety and security. However, in terms of nuclear safety, it is important to recognize potential risks especially when considering a controlled site where largely arbitrary values of risk criterion have been placed at boundaries of those sites (typically 1 in 10^4/year). There is a sense that because of these difficulties, there is a shift in how we think about risk. In the United States, the Department of Energy has published a new interpretation of high-level waste focusing on the characteristics of the waste, not on how it originates (US DoE, 2019). The purpose is essentially to reduce the need for deep underground disposal. Past policy prescriptions for disposal arose from a policy based solely on the radioactivity the waste contains. This is perhaps not a surprising approach in a systems-thinking world, but in the world of public opinion, it is a challenge because the public perception of risk is substantially higher than the real objective risk. Such realities give rise to an interesting perspective suggesting new approaches will become available in regard to security and safeguards as technology progresses, but a barrier will be the public perception of risk.

The approaches taken by nations can be seen to vary, especially in decommissioning. One can observe the recent decision by the US government to consolidate waste fuel under the proposed Nuclear Waste Policy Amendments Act, 2019 into a single federal facility. This action seeks not only to limit the risk from seismic activity but also to promote a permanent repository at Yucca Mountain. The issue of safety and safeguards perhaps leads to a focal point to any terrorist activity such as perhaps the UK plant at Sellafield. At one level, this is again 'concentrate and control' as opposed to the more discredited "dilute and disperse" strategy. Lack of consideration of the "dilute and disperse" alternative can miss the fact that the negative consequences might be very limited, such as the very modest harm that would be expected to occur from the release of ^{14}C if all irradiated nuclear graphite moderator components were to be disposed of through incineration.

It is also interesting to recall that the United Kingdom back in the 1980s had developed plans for a deep rock laboratory in preparation for a deep repository near Sellafield, but some decades on, we are in reality no further forward with no potential site selected. So, from a UK perspective, there are no fundamental issues that would prevent the country from a safety case and thus a future facility for geological disposal. However, there are some concerns, but nevertheless as this procrastination continues, nuclear safety gives rise to perhaps a higher objective risk than the public perception appreciates.

The Fukushima incident presented a challenge to how both public and expert communities considered the challenges and the perceived risks from the hazards. The events that led to the accident were, we suggest, insufficiently considered, which in turn meant the foreseeable risks were not fully or appropriately mitigated. Consequently, it is not inconceivable that a paradigm shift in reactor safety analysis might ultimately result; indeed, some would say that this has already occurred with the invocation of the concept (borrowed from banking after the financial crisis of 2008) of the "stress test." We would suggest that while the stress test may indeed have been a responsive shift, it was not the best thought through policy response that could have been developed (see www.nrefs.org). It is perhaps a perverse consequence of Fukushima that electricity prices increased and greater use of fossil fuels have led to more deaths following the Fukushima accident in March 2011 than the subsequent evacuation from the area surrounding the nuclear power plant as shown in a study by Neidell et al. (2019). It is also useful to remind ourselves that no deaths were recorded as a direct and immediate result of the nuclear accident, but the decision to suspend nuclear power generation in response to it has contributed to loss of life by increased imports of fossil fuels with increases in electricity prices by as much as 38% in some parts of Japan. The consequence of which was seen to be a decrease in electricity consumption during peak periods. The authors estimate the additional deaths (circa 1000), it may be argued, are an illustration on how decisions related to nuclear safety can create perverse consequences.

4. Summary remarks

As noted earlier, civil nuclear applications range far more widely than just the production of electricity: from the early luminance of watch hands in the early 1900s, through the nuclear-powered pacemakers of the early 1970s (Reuters, 2007 reported some of which are still operating after a lifetime!) to more recent medical therapies and diagnostic tools. International collaboration has featured strongly in the development of civil nuclear technologies, and the role of the IAEA has been a tremendous success in this regard. The prospect of growth, arising from nuclear renaissance of a move into new applications (such as VHTR process heat applications), does pose new risks that were not as evident in the early days. The risks-related issues include emergency preparedness and, as seen in the case of the Chernobyl accident of 1986, a need to manage risks that cross international borders. Finally, there is the latent risk associated with an increase in the private activity that once was confined to state actors.

We close with the suggestion that the long-awaited western Nuclear Renaissance might still occur, but not as once envisaged, being no longer focused narrowly on electricity generation. The global drive toward deep decarbonization could provide a major impetus to process heat applications of nuclear energy. This will prompt technological innovation and present new challenges and opportunities. It will not fundamentally alter, however, the current framing of security, safeguards, and safety. That framing of the issues presented by nuclear fission will remain appropriate as nuclear fission finds application in new directions. We also note that while public perception and expert technocracy may perceive different things about nuclear energy, both aspects need to be appreciated by policy-makers in a democracy. For nuclear energy to have a sustainable future, it must be safe and secure, and equally importantly, it must be perceived as such. Those tasked with developing future nonelectrical applications of nuclear technology should keep such realities in mind.

Acknowledgments

The authors are grateful for the support given by The Open University.

References

ACS, 2010. Detecting Fake Wine Vintages: It's an (Atomic) Blast. https://www.acs.org/content/acs/en/pressroom/newsreleases/2010/march/detecting-fake-wine-vintages-its-an-atomic-blast.html.

Ashley, S.F., Nuttall, W.J., Parks, G.T., Worrall, A., 2012. On the proliferation resistance of thorium-uranium nuclear fuel. In: Presented at the 2012 UK PONI Annual Conference ' Nuclear Stability: From the Cuban Crisis to the Energy Crisis'.

Brown, A., Glaser, A., 2016. On the origins and significance of the limit demarcating low-enriched uranium from highly enriched uranium. Science and Global Security 24 (2), 131−137.

Cholewinski, M., Tomkow, L., 2017. Assessment of district heating needs and technical possibilities in the vicinity of a future polish nuclear power station. Interdisciplinary Journal of Engineering Sciences V (1).

Csik, B.J., Kupitz, J., 1997. Nuclear power applications: supplying heat for homes and industries. IAEA Bulletin 39 (2), 21−25.

Energy Transitions Commission, 2020. Reaching Net-Zero Carbon Emissions: Mission Impossible. http://www.energy-transitions.org/mission-possible.

EU, 2016. European Parliament Briefing: Chernobyl 30 Years on - Environmental and Health Effects.

FISA, 2019. In: EU/Euratom Conference on Fission Safety of Reactor Systems, Pitesti, June 2019.

Forsberg, C., Brick, S., Haratyk, G., 2018. Coupling heat storage to nuclear reactors for variable electricity output with baseload reactor operation. The Electricity Journal 31 (3), 23−31. https://doi.org/10.1016/j.tej.2018.03.008.

GIF IV International Forum, 2020. https://www.gen-4.org/gif/jcms/c_9260/public.

Glaser, A., 2005. About the enrichment limit for research reactor conversion: why 20%?. In: The 27th International Meeting on Reduced Enrichment for Research and Test Reactors (RERTR), November 6−10, 2005, Boston Massachusetts.

Guth, S., 2018. Oasis of the future: the nuclear city of Shevchenko/Aqtau, 1959−2019. Jahrbücher für Geschichte Osteuropas 66 (1), 93−123.

Hodgins, G.W., 2009. Measuring Atomic Bomb-Derived 14C Levels in Human Remains to Determine Year of Birth and/or Year of Death. U.S. Department of Justice Report 227839.

Horak, W.C., 1997. Co-generation in the former Soviet Union. In: IECEC-97 Proceedings of the Thirty-Second Intersociety Energy Conversion Engineering Conference (Cat. No.97CH6203), Honolulu, HI, USA, vol. 2, pp. 1241−1245.

IAEA, 2006. Chernobyl's Legacy: Health, Environmental and Socio-Economic Impacts and Recommendations to the Governments of Belarus, the Russian Federation and Ukraine. https://www.iaea.org/sites/default/files/chernobyl.pdf.

IAEA, 2008. In: Chernobyl: Looking Back to Go Forward - 2005, Conference Proceedings, STI/PUB/1312. https://www-pub.iaea.org/MTCD/publications/PDF/Pub1312_web.pdf.

IEA, 2017. IEA Greenhouse Gas R&D Programme Technical Review 2017 − TR3. Reference Data and Supporting Literature Reviews for SMR Based Hydrogen Production With CCS.

IAEA, 2010. Progress in Radioactive Graphite Waste Management. IAEA-TECDOC-1647.

IAEA, 2015. Manual of Good Practice in Food Irradiation, Technical Report 481. IAEA Vienna. ISBN: 978-92-0-105215-5.

IAEA, 2016a. IAEA Nuclear Energy Series No. NP-T-4.3 Industrial Applications of Nuclear Energy. ISBN: 978-92-0-101417-7.

IAEA, 2016b. IAEA Safety Standards Series No. SSR-2/1 (Rev. 1). Safety of Nuclear Power Plants: Design Specific Safety Requirements. ISBN: 978-92-0-109315-8.

Jonter, T., 2010. The Swedish plans to acquire nuclear weapons, 1945−1968: an analysis of the technical preparations. Science and Global Security 18 (2), 61−86. https://doi.org/10.1080/08929882.2010.486722.

Khan, S.U., Khan, S.U., Danish, S.N., Orfi, J., Rana, U.A., Haider, S., 2018. Nuclear energy powered Seawater desalination, chapter 6. In: Veera Gnaneswar, G. (Ed.), Renewable

Energy Powered Desalination Handbook: Application and Thermodynamics, pp. 225–264. https://doi.org/10.1016/B978-0-12-815244-7.00006-4. ISBN: 0128152443.

Kugeler, K., Zhang, Z., 2019. Modular High-Temperature Gas-Cooled Reactor Power Plant. Springer-Verlag GmbH, Germany. https://doi.org/10.1007/978-3-662-57712-7. ISBN: 978-3-662-57710-3.

Mathers, D., 2019. UK Progress in Enabling SMRs and Advanced Nuclear Technology. IAEA TWG-SMR, Vienna. https://nucleus.iaea.org/sites/htgr-kb/twg-smr/Documents/TWG-2_2019/D10_A%20UK%20framcework%20for%20Advance%20Nuclear%20Technologies%20IAEA%20TWG-SMR%20draft.pdf.

McGuire, M.A., Neighbour, G.B., Price, R., 2010. Should subjective risk be taken into account in the design process for graphite disposal? In: Neighbour, G.B. (Ed.), Securing the Safe Performance of Graphite Reactor Cores. Royal Society of Chemistry. https://doi.org/10.1039/9781847559135.

NIRAB, 2019. NIRAB Annual Report 2018/19 - Clean Growth Through Innovation — the Need for Urgent Action, p. 15. http://www.nirab.org.uk/our-work/annual-reports/.

Neidell, M., Uchida, S., Veronesi, M., 2019. Be Cautious With the Precautionary Principle: Evidence From Fukushima Daiichi Nuclear Accident. I Z A Institute of Labor Economics Discussion Paper Series IZA DP No. 12687.

Nuclear Industry Council, 2018. The Nuclear Sector Deal. Nuclear Industry Council Proposals to Government for a Sector Deal. Available from: https://assets.publishing.service.gov.uk/government/uploads/system/uploads/attachment_data/file/665473/The_Nuclear_Sector_Deal_171206.pdf.

Nuttall, W.J., Bakenne, A.T., 2020. Fossil Fuel Hydrogen-Technical, Economic and Environmental Potential. Springer. https://doi.org/10.1007/978-3-030-30908-4. ISBN: 978-3-030-30907-7.

Nuttall, W.J., Glowacki, B.A., Clarke, R.H., 2005. A Trip to Fusion Island, the Engineer. https://www.theengineer.co.uk/a-trip-to-fusion-island/.

Nuttall, W.J., Glowacki, B.A., July 2008. Viewpoint: Fusion Island, Nuclear Engineering International, pp. 38–41.

Nuttall, W.J., Storey, P., 2014. Technology and policy issues relating to future developments in research and radioisotope production reactors. Progress in Nuclear Energy 77, 201–213.

ONR, 2014. Safety Assessment Principles for Nuclear Facilities. ONR CM9 Ref 2019/367414 (Revision 1, Jan 2020). http://www.onr.org.uk/saps/saps2014.pdf.

Paul, T.V., 2000. Power Versus Prudence: Why Nations Forgo Nuclear Weapons. McGill-Queen's Press - MQUP.

Peakman, A., Owen, H., Abram, T., 2019. The core design of a small modular pressurised water reactor for commercial marine propulsion. Progress in Nuclear Energy 113, 175–185. https://doi.org/10.1016/j.pnucene.2018.12.019.

Rachkov, V.I., Kalyakin, S.G., Kukharchuk, O.F., et al., 2014. From the first nuclear power plant to fourth-generation nuclear power installations [on the 60th anniversary of the World's first nuclear power plant]. Thermal Engineering 61, 327–336. https://doi.org/10.1134/S0040601514050073.

Reuters, 2007. Nuclear Pacemaker Still Good after 34 Years. https://uk.reuters.com/article/health-heart-pacemaker-dc/nuclear-pacemaker-still-energized-after-34-years-idUKN1960427320071219.

Rosatom, 2020. https://www.rosatom.ru/en/rosatom-group/non-nuclear-equipment-manufacturing/desalination/.

U.S. DoE, June 2019. Press Release - Independent Reports Supporting a Risk-Based Approach to Radioactive Waste Management.

Vital Energi, 2020. https://www.vitalenergi.co.uk/blog/second-phase-of-stoke-on-trents-heat-network-gets-under-way/.

White, I.F., Smith, G.M., Saunders, L.J., Kaye, C.J., Martin, T.J., Clarke, G.H., Wakerley, M.W., 1985. Assessment of Management Modes for Graphite From Reactor Decommissioning. EEC Report EUR9232 (Contract DE-D-001-UK).

Wilson, A., Nuttall, W.J., Glowacki, B.A., 2020. Techno-economic Study of Output-Flexible Light Water Nuclear Reactor Systems With Cryogenic Energy Storage, EPRG Working Paper 2001, Posted on 7 January 2020. https://www.eprg.group.cam.ac.uk/eprg-working-paper-2001/.

WNA, 2020. http://www.world-nuclear.org/information-library/non-power-nuclear-applications/radioisotopes-research/research-reactors.aspx.

WNN, 2020. https://world-nuclear-news.org/Articles/Australian-isotope-facility-to-begin-full-scale-op.

Zverev, D.L., Fadeev, Y.P., Pakhomov, A.N., et al., 2019. Nuclear power plants for the icebreaker fleet and power generation in the Arctic region: development experience and future prospects. Atomic Energy 125, 359–364. https://doi.org/10.1007/s10512-019-00494-5.

Radiation hazards from the nuclear fuel cycle

Man-Sung Yim, ScD, PhD, MS, SM

Professor, Nuclear and Quantum Engineering, Korea Advanced Institute of Science and Technology, Daejeon, South Korea

1. Basic concepts of radiation hazards

Radiation is the energy released from an unstable atom in an effort to become less unstable or stable. Radiation is present throughout the nuclear fuel cycle in the form of various particles such as photons, neutrons, β-rays (electrons), and α-particles. The photons include X-ray or γ-ray as electromagnetic radiation with a specific wavelength. X-ray originates from orbital electrons and involves the movement of electrons, while γ-ray originates from atomic nucleus. α-Particle is identical to the nucleus of the helium atom consisting of two protons and two neutrons bound together into a particle. These particles are ionizing radiation as they carry enough kinetic energy to cause ionization (i.e., removal of orbital electron) of the target atom. When the energy of radiation is not large enough to produce ionization of an atom with which the radiation interacts, the radiation is called nonionizing radiation. In the case of radioactive decay of unstable atoms, the released particles are all ionizing radiation.

Ionization of an atom or molecule results in the production of an ion pair, i.e., an electron removed from the target atom or molecule and the parent atom or molecule to which it was originally attached.

Such production of ionization is an outcome of direct or indirect interactions of the incoming radiation with an atom or molecule. Electrically charged particles (e.g., electrons or α-particles) cause ionization through direct charge-to-charge interactions with the orbital electrons of the target atom. Charged particles are therefore called directly ionizing particles. Uncharged particles such as photons or neutrons often go through interactions with the atomic nucleus that subsequently result in the release of electrons. Thus uncharged particles are called indirectly ionizing particles.

In the case of directly ionizing particles, the radioactive particles continuously lose their energy in the target material through impulses exerted at a distance as electric forces between the charged particles and orbital electrons. Such charge versus charge interactions are called coulombic interactions. In contrast, the interaction of indirectly ionizing particles requires hitting the "target" (including a "close encounter" with the target, but not necessarily physically contacting one). Therefore the interactions are "by chance" and occur when the indirectly ionizing radiation approaches close enough to the target. The amount of energy transferred to the target material in a given unit distance is much larger with directly ionizing particles because of the large number of interactions involved than that with indirectly ionizing particles. Also the tracks of ionization are much more densely populated with directly ionizing particles in comparison to indirectly ionizing particles. This in turn means that the distance of the radioactive particle movement in the target is relatively short with directly ionizing particles, whereas indirectly ionizing particles can penetrate the target such as a structure or the human body.

Accordingly, protection against directly ionizing particles (e.g., α- and β-particles) is easier in comparison to indirectly ionizing particles. The typical distance traveled by electrons in water is up to a few millimeters. In the case of α-particles, the distance is much less than a millimeter. Thus shielding against electrons and α-particles can be readily achieved. For example, α-particles can be completely stopped by a sheet of paper. Electrons can be stopped by a sheer of plastic. In this case, the main hazards from directly ionizing radiation are through the intake of the corresponding particles into the human body by ingestion or inhalation. In contrast, indirectly ionizing particles can travel much longer distances within a target or penetrate the target and thus becomes a source of hazard through external irradiation. Shielding against γ-rays is often provided by the use of high-density heavy materials such as lead. For shielding against neutrons, water, plastic, or concrete is typically used. Shielding against indirectly ionizing particles requires a detailed analysis to achieve the necessary degree of protection.

Examples of major radionuclides present in the nuclear fuel cycle are shown in Table 3.1. As seen in Table 3.1, these radionuclides are sources of different ionizing radiation such as α-particles, β-particles, and γ-rays.

As the energy of ionizing radiation is transferred from the incident particle and deposited into the target medium, a breakup of a molecule or other molecular changes in the interacting medium take place. This becomes the basis of radiation hazards (including health effects of radiation). This the resulting energy deposition would be directly related to the damage in the target. Thus radiation hazards arise from the energy deposited from ionizing radiation in the biological system. Such deposited energy is termed "dose." Dose represents the energy deposited in a mass and is used as the quantity to represent potential damage associated with radiation exposure.

There are a number of derived definitions of dose depending on the purpose. The first quantity to be defined is the absorbed dose. The absorbed dose specifies the amount of energy absorbed per unit mass of material from radiation interaction with the matter. More specifically, it is the average energy imparted to the mass contained in an incremental volume of a specified material. The unit for absorbed dose is gray, written as Gy. 1 Gy represents 1 J of energy deposited in 1 kg of mass. 1 Gy is also equal to 100 rad; rad (radiation absorbed dose) is another (older) unit of absorbed dose. 1 rad is equal to 100 ergs per gram (1 erg $= 10^{-7}$ J).

Table 3.1 Examples of key radionuclides and their characteristics in nuclear fuel cycles.

Nuclide	Half-life	Radiation	Particle energy (MeV)	Production	Presence
^{3}H	12.32 years	β-ray	0.0057 (average)	Neutron activation, ^{2}H(n,γ)^{3}H, ^{10}B(n,^{8}Be)^{3}H, ^{6}Li(n,α)^{3}H, ^{7}Li(n,nα)^{3}H, or fission product	Nuclear reactor operation
^{14}C	5730 years	β-ray	0.0495 (average)	Neutron activation, ^{14}N(n,p)^{14}C, ^{17}O(n,α)^{14}C	Nuclear reactor operation
^{60}Co	5.271 years	β-ray and γ-ray	β: 0.0958 (average) γ: 1.3325 and 1.1732	Activation, ^{59}Co(n,γ)^{60}Co	Nuclear reactor operation
^{85}Kr	10.72 years	β-ray	β: 0.251 (average)	Fission product	Reprocessing
^{90}Sr	29.1 years	β-ray	0.196 (average)	Fission product	Spent fuel disposal
^{90}Y	2.67 days	β-ray and γ-ray	β: 0.935 (average) γ: 2.286	Decay product of ^{90}Sr	Spent fuel disposal
^{99}Tc	2.13 × 10^5 years	β-ray and γ-ray	β: 0.0846 (average) γ: 0.0897	Fission product	Spent fuel disposal
^{129}I	1.57 × 10^7 years	β-ray and γ-ray	β: 0.050 (average) γ: 0.0396	Fission product	Spent fuel disposal
^{131}I	8.0 days	β-ray and γ-ray	β: 0.192 (average) γ: 0.3645 (81.5%); 0.6370 (7.16%); 0.2843 (6.12%); 0.0802 (2.62%)	Fission product	Nuclear reactor accident
^{133}Xe	5.25 days	β-ray and γ-ray	β: 0.101 (average) γ: 0.081 (37.4%); 0.031 (39.1%); 0.035 (9.11%)	Fission product	Nuclear reactor operation & reprocessing
^{137}Cs	30.08 years	β-ray	β: 0.173 (average)	Fission product	Reactor accident, spent fuel disposal

Continued

Table 3.1 Examples of key radionuclides and their characteristics in nuclear fuel cycles.—*cont'd*

Nuclide	Half-life	Radiation	Particle energy (MeV)	Production	Presence
^{137m}Ba	2.552 months	γ-ray	0.6617 (89.8%)	Decay product of ^{137}Cs	Reactor accident, spent fuel disposal
^{222}Rn	3.82 days	α-ray	5.4895	Part of ^{238}U decay chain	U mining
^{226}Ra	1600 years	α-ray and γ-ray	α: 4.784 γ: 0.186 (3.28%)	Part of ^{238}U decay chain	
^{232}Th	1.40×10^{10} years	α-ray	4.0123 (78.2%), 3.9472 (21.7%)	Naturally existing	
^{238}U	4.468×0^{9} years	α-ray	4.198 (79.0%), 4.151 (20.9%)	Naturally existing	Front-end fuel cycle
^{239}Pu	24110 years	α-ray	5.1566 (70.77%), 5.1443 (17.11%), 5.1055 (11.94%)	^{238}U(n,γ)^{239}U (β$^-$ decay) $\rightarrow$ ^{239}Np (β$^-$ decay) $\rightarrow$ ^{239}Pu	Spent fuel

The equivalent dose is a modified dose concept introduced for the consideration of biological effect. While the absorbed dose does not take into account the differences in biological effectiveness of different particles, the equivalent dose is obtained after multiplying the absorbed dose by radiation weighting factor (ICRP, 2007). Radiation weighting factor reflects in the differences in biological effectiveness depending on the type and energy of the radiation. The unit for the equivalent dose is sievert, abbreviated as Sv. 1 Sv is equal to 1 Gy times the radiation weighting factor of 1, corresponding to 1 J energy deposition per 1 kg by a radiation with radiation weighting factor of 1. Another (older but still widely used) unit for equivalent dose is rem, which is equal to 0.01 Sv. Thus 1 rem = 1 rad * Q, 1 Sv = 1 Gy * Q. The radiation weighting factor for electrons, X-rays, and γ-rays is 1.0. For α-particles, the radiation weighting factor is 20. For neutrons, the radiation weighting factor varies between 2.5 and 20 as a function of neutron energy.

Once the dose to various organs of an exposed person is defined by using the concept of equivalent dose, these organ-specific doses are combined to represent the whole-body dose. Such combined total dose to the whole body is termed effective dose. As different organs may show different sensitivity to radiation, the differences in radiosensitivity of organs are taken into account by using the organ weighting factor. The organ weighting factor is called tissue weighting factor. The tissue weighting factor is derived from experimental observations from the atomic bomb survivors study by observing how many organ-specific cancers appear per given whole-body dose to the exposed individuals. The tissue weighting factors currently in use are as follows: 0.12 for bone marrow, colon, lung, stomach, and breast; 0.08 for gonads; 0.04 for bladder, liver, esophagus, and thyroid; and 0.01 for skin, bone surface, and salivary glands. The unit for the effective dose is the same as the equivalent dose.

2. Overview of the nuclear fuel cycle

The nuclear fuel cycle begins with extracting uranium from the nature and ends with disposition of the residual uranium and any byproduct materials derived from the uranium as waste. With the exception of a few nuclear reactor types (e.g., heavy water reactor [HWR] or gas cooled reactor [GCR]), most nuclear reactors use fuels with fissile ^{235}U at an isotopic concentration greater than the natural uranium level (i.e., 0.711 wt%). Therefore increasing ^{235}U isotope content is necessary as part of the nuclear fuel cycle. Such work is called (uranium) enrichment. The front end of the fuel cycle includes mining of uranium ore, milling of uranium ore to produce U_3O_8 (yellowcake), purification of U_3O_8 and conversion to UF_6 as a preparatory step for enrichment, enrichment of ^{235}U as UF_6 (to the required ^{235}U level), and fuel fabrication. The fabricated nuclear fuel is loaded to nuclear reactor for power generation. The back end of the fuel cycle covers used-fuel management and other follow-on activities until used fuels are permanently disposed of. These activities include on-site used-fuel storage, used-fuel shipment, away-from-reactor storage, and final

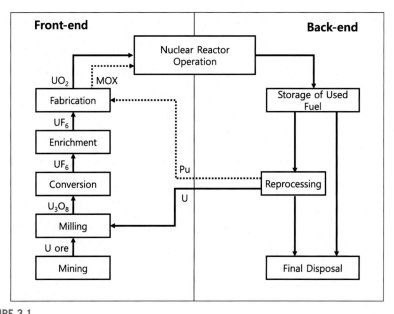

FIGURE 3.1

Steps of the nuclear fuel cycle.

disposition. Key steps of the nuclear fuel cycle in the front end and back end are shown in Fig. 3.1, where possible recycling of uranium and plutonium is also depicted.

2.1 Mining

Uranium mining involves removing large quantities of rocks and soils with elevated concentration of uranium from underground. The rocks and soils in uranium mines have the concentration of uranium up to several thousand times the average concentrations of uranium in the earth's crust with about 0.15%—0.3% of uranium content. The mined natural U is composed of two main isotopes: ^{235}U (0.72 wt %) and ^{238}U (99.27 wt %). As a natural ore material, uranium-containing rocks also contain other ubiquitous metals in the earth's crust, such as ^{232}Th, ^{234}Th, and ^{226}Ra, which need to be removed for the use of uranium as fuel.

Uranium mining is performed by using three different techniques, i.e., underground mining, open-pit (surface) mining, or solution, in situ leaching (ISL) mining. Underground mining is used for uranium deposits located deeply or covered by strata of hard rock, while open pit mining is used when uranium ore is within a few hundred meters from the surface and when overburden can be removed without excessive use of blasting. Underground mining involves higher occupational radiation dose to the miners, while open pit mining comes with the disadvantage of greater environmental impacts due to the larger surface area disturbed by the mining activity. Solution mining is based on solubilizing the uranium in an underground ore body by injecting leaching solutions. The dissolved uranium solution is pumped out followed by uranium recovery operations. This technique produces much less amount of mining waste compared with the physical extraction technique, which requires disposal but comes with a higher cost. Applicability of the approach is limited to uranium ore body in a generally horizontal bed underlain by a relatively impermeable stratum. Use of a particular method depends on the nature of the ore body and safety and economic considerations.

2.2 Milling

As the level of uranium concentration in mined ore is low at around 0.15%–0.3%, it is necessary to raise the uranium content before further processing of the mined ore. Milling is to remove other constituents in the ore and to obtain a highly concentrated uranium mass (typically up to 85–90 wt %). For such purposes, a combination of physical and chemical processes is used. The product of uranium milling is called "yellowcake (U_3O_8)."

In milling, uranium is first crushed and grinded to a suitable (i.e., small) size to increase the surface area. Uranium in the crushed particles is dissolved by using an acid and combined with a solvent to be extracted into organic phase. Uranium in the organic phase is then stripped back to aqueous phase. This step is called solvent extraction. This is followed by additional steps of chemical precipitation, calcination, pulverization, and packaging to produce the product, the yellowcake.

2.3 Conversion

The yellowcake produced from uranium milling contains impurities such as boron, cadmium, and rare earth elements. Such impurities need to be removed, as they have sizable neutron-absorption effects in the fuel. Removal of impurities from yellowcake is called refining. Thus it is the purification process of yellowcake. The process is again based on the solvent extraction technique. The yellowcake is dissolved in nitric acid and uranium is extracted by using an organic solvent (called tributyl phosphate [TBP]) as the organic phase. The extracted organic complex is then treated to back-extract uranium as uranyl nitrate solution.

The purified uranyl nitrate solution is then evaporated to a uranyl nitrate molten salt, which is subsequently denitrified and reduced to UO_2. The product, UO_2, can be used as the fuel if enrichment of uranium is not required. If enrichment is needed, UO_2 needs to be converted into uranium hexafluoride [UF_6], a gaseous form of uranium.

UF_6 is the only known compound of uranium that is suitable for the gaseous diffusion and other gas-based enrichment processes as the only sufficiently volatile compound of uranium. Also fluorine as a stable element has only one isotope ^{19}F, so that the difference in molecular weights of UF_6 molecule is due only to the difference in weights of the uranium isotopes. The product UF_6 is shipped to an enrichment facility in metal containers as a solid.

2.4 Enrichment

Uranium in the form of UF_6 goes through enrichment. The techniques of enrichment include gaseous diffusion, gas centrifuge, and other techniques such as aerodynamic separation, electromagnetic separation, laser separation, and chemical separation. Currently, large-scale commercial operations of uranium enrichment are based on either gaseous diffusion or gas centrifuge.

In gaseous diffusion, gaseous UF_6 will move through a porous wall from a region of higher pressure to a region of lower pressure. As the energies of gas molecules of differing masses are the same at thermal equilibrium, lighter molecules have a slightly higher speed and move slightly faster than the heavier ones, leading to the separation of $^{235}UF_6$ from $^{238}UF_6$.

In gas centrifuge, the UF_6 gas rotates inside the cylinder subject to centrifugal acceleration. As the gas molecules rotate at the same speed, the energy of heavier $^{238}UF_6$ molecules is higher. Accordingly, the heavier $^{238}UF_6$ molecules are forced toward the outside of the cylinder, whereas the lighter $^{235}UF_6$ molecules tend to collect closer to the center, leading to the separation of ^{235}U and ^{238}U.

2.5 Fuel fabrication

Fuel fabrication is the process in which the enriched UF_6 gas is converted into the fuel form of choice, followed by manufacturing of fuel rods and assembly. This step includes fuel pellet preparation, fuel rod fabrication, and fuel assembly fabrication. While the last two stages are mostly mechanical processes, fuel pellet preparation includes conversion of enriched UF_6 to UO_2 and production of UO_2 pellet. UO_2 is the most common form of fuel for commercial nuclear power operation including the use in pressurized water reactor (PWR), boiling water reactor (BWR), HWR, light water graphite reactor (LWGR), and fast breeder reactor (FBR).

For UF_6 conversion into UO_2, the solid UF_6 is sublimed to its gaseous form and is reacted with superheated steam using a rotary kiln for conversion to UO_2 powder. The UO_2 powder is milled and pressed into pellets. The pressed pellets are sintered to be converted into a high-density ceramic with the target porosity of about 5% (i.e., up to 95% of theoretic density). The sintered pellets are outgassed and ground for the final finished product.

The finished pellets are stacked up inside cylindric zircaloy tube, known as classing, and sealed up with welding after filling the voids with inert (e.g., helium) gas to produce a fuel rod. The finished rods are placed in the cage assembly to form a fuel assembly. The position of each fuel rod is fixed in the prefabricated framework structures that hold the rods in a precisely defined grid arrangement. These finished fuel assemblies are about 4–4.5 m long and weighing about 660 or 320 kg for a PWR or BWR assembly, respectively.

2.6 Reactor operation

The reactor operations stage is where the fabricated fuel assemblies are loaded into nuclear reactor for the irradiation of nuclear fuels to cause fission of uranium. With each fission reaction producing about 200 MeV as thermal energy, the energy generated is utilized for steam production and electricity generation.

At the beginning of reactor operation, the reactor must contain excess fissile materials to maintain criticality for the rest of the operating cycle. This means that an extra control mechanism must be provided to maintain the reactor critical during the entire operating life while containing excess fissile materials. This is done by controlling the level of neutron poisons such as soluble boron or burnable poison rods in the core. As the fissile materials are depleted due to fission, the level of neutron poisons is reduced to keep the reactor exactly at the critical condition. When the reactor comes to a point where maintaining criticality is no longer possible because of the depletion of fissile materials even with the removal of neutron poisons, the fuel reaches the limit of its use and needs to be replaced with fresh fuel. At this point the reactor has reached the so-called cycle length, which is the designed irradiation period of the fuel. Then the fuel is discharged from the reactor and becomes used nuclear fuel. The cycle length typically ranges between 3 and 5 years. Therefore the fuel, once loaded, stays in the core for 3–5 years depending on the design of the operating cycle while going through reshuffling in the core. Reshuffling is to move the fuel assemblies into the locations where the remaining fissile contents of the assemblies are best utilized while maintaining the core power distributions as uniform as possible. The higher the initial content of the fissile ^{235}U in the fuel, the longer the cycle length is. Used nuclear fuels are typically called "spent nuclear fuels." Hereafter, used fuel will be referred to as spent fuel. These spent fuels still contain fissile materials (e.g., ^{235}U, ^{239}Pu). Therefore employing a scheme to recycle the fissile material back to reactor operation for further energy generation could be considered.

The changes in the composition of fuel depend on how much energy is produced by fission in the fuel. The amount of energy produced per unit mass of fuel is a useful measure of fuel utilization and is defined as "burnup." The burnup is determined by the level of power produced and the length of fuel irradiation (reactor operation) and is represented by megawatt days (MWD) per unit mass of fuel used (e.g., ton). Thus burnup can be computed by multiplying the thermal power of the nuclear reactor by the time of operation and dividing by the mass of the initial fuel loading. Note that here metric ton is used to represent the mass of the fuel by using the term MTHM (metric ton heavy metal), the total mass of the heavy metals (i.e., uranium) used in the fuel.

Concern of radiation hazards during reactor operation comes from the presence of large quantities of radioactive materials in the reactor. The radioactive materials can be discharged during normal operation through airborne and waterborne pathways. Such discharges can affect the human food chain, thus presenting hazard to the public through the consumption of contaminated food or water. Release of radioactive materials in large quantities can also occur if a severe nuclear accident, such as the Chernobyl or Fukushima accident, takes place.

2.7 Storage of spent fuel

In a typical light water reactor, about a quarter to a third of the total fuel load is removed from the core every 12–24 months through refueling. Thus a certain portion of the loaded fuels is always replaced with fresh fuels becoming spent (or "used") fuel. These spent fuels are stored on-site until they are shipped away from the reactor for storage, reprocessing, or permanent disposal.

Spent fuel storage can be classified into two categories: water pool (wet) and dry storage. Virtually all nuclear power reactors have an on-site water-filled pool for the storage of spent fuel. Water, as a storage medium, is inexpensive, is good for neutron shielding, and provides excellent cooling capacity through natural circulation. Spent-fuel storage facility can also be built away from the reactor site typically by using dry storage to meet the additional storage space needed. In both wet and dry storage technologies, the spent fuel must be appropriately protected through implementation of shielding, cooling, and criticality control.

For wet storage, the discharged nuclear fuel from nuclear reactor is moved to the storage pool by using automated handling systems. Water allows clear visibility for inspection and handling. The pools are rectangular in horizontal cross section and 12–13 m deep. The pool is typically surrounded by a steel-lined concrete wall (>1 m thick) as a leak-tight structure. The storage zone in the pool is located near the bottom of the pool (about 4 m high). In the storage zone, the spent fuel assemblies are stored within the storage racks. The storage racks are to provide spacing for coolant flow and to keep the fuel in controlled positions for physical protection and for ease of tracking and rearrangement. They are placed at the bottom of the pool in a lattice array made of aluminum steel. For criticality control, the spacing of the spent fuel assemblies in the storage racks is adjusted and boric acid is added to the pool water. Above the spent fuel storage zone, more than 7 m of water is used for shielding.

Dry storage is not appropriate for spent fuel with high heat loads, such as freshly discharged spent fuels, because of the low efficiency in cooling with air. However, for spent fuels with low heat loads (e.g., spent fuels cooled for at least 1 year in the spent fuel pool), dry storage can be appropriate. In dry storage the spent fuel is stored in a shielded container outside the reactor containment building. The storage can be in casks, a small modular structure, or a large building. It can be a surface facility or an underground facility. Dry storage offers flexibility and can be tailored to meet the needs of a specific site. In the case of using casks or a small modular structure, expansion of the storage capacity can be readily achieved in very small increments and on relatively short notice once the required packaging facilities are available. Although continuing surveillance and maintenance of the facilities are needed, maintenance requirement for dry storage is lower than that for wet storage.

2.8 Transportation of spent fuel

Transportation provides an essential link between different phases of spent-fuel management. Transport of spent fuel is typically done by rail or truck or by the sea. In all modes of transport, special casks are used, with their design and size being different depending on the mode of transportation. The goals of shipping cask design are to provide physical containment, radiation shielding, heat removal, criticality protection, and theft protection.

The shipping casks are specially manufactured containers with various components designed to meet the design requirements. The casks consist of a large cylindric steel vessel, shielding layers, fuel baskets, a bolt-down lid, and impact limiters. Fuel assemblies are placed inside fuel support baskets within the inner cavity designed to accommodate different types of assemblies. The baskets provide mechanical support structure. Boron can be included inside the structural material for criticality control.

Two layers of shielding are provided inside the vessel of fuel baskets. The first shield layer surrounding the fuel cavity is the γ shield. The γ shield is made of lead, depleted uranium, or steel. The second layer is the neutron shield. Neutron shield consists usually of boron in a mixture of water and ethylene glycol jacket or a specially prepared polyester resin compound. The two layers are sandwiched between steel shells and the outer side of the cask is solid steel. A massive bolt-down lid is used for the closure of the casks. Impact limiters are placed at each end of the vessel to limit the impact of collision accidents. Cooling of the casks is primarily through passive heat transfer through the cask walls. On some casks, cooling fins are provided to increase the heat transfer area.

2.9 Reprocessing of spent fuel

Reprocessing is a series of physical and chemical operations to separate the elements of interest or concern from spent fuel. The most widely used method today is the PUREX process that separates U and Pu. In the process the mechanical barrier of spent fuel is broken and the fuel rod is dissolved in nitric acid. The resulting mix of dissolved fuel rod is contacted by an organic solvent (i.e., TBP) that selectively forms bonding with uranium and plutonium. This results in the separation of U and Pu in the organic phase while all other constituents of fuel remain in the nitric acid aqueous phase. The remaining constituents form high-level radioactive waste (HLW). The U and Pu in the organic phase are mechanically removed and chemically treated for the separation of Pu and U. Pu isotopes, once separated, can be recycled into nuclear reactors for energy generation through mixed oxide fuel fabrication. The separated uranium can also be recycled for fuel fabrication or can be disposed of as low-level waste. The waste products of reprocessing, such as HLW, can be converted into stable waste forms typically as glass that are suitable for safe, long-term storage.

2.10 Permanent disposal of spent fuel

For the permanent disposal of spent fuel or the HLW from reprocessing, various options have been considered. These options include disposal in deep seabed, disposal inside icebergs, disposal in very deep holes, injection of liquid into deep wells, geologic disposal on small uninhabited islands, disposal in mined geologic repositories, and sending the waste into the space. These methods are basically to move the waste to a medium that is stable and far away from people and to keep the waste within such medium for very long periods.

Out of those options, disposal of nuclear waste in mined geologic repositories has been selected as the method of choice by several countries in the 1980s. Today, international consensus exists on the use of mined geologic repositories for high-level waste disposal. Operation of licensed geologic repositories is expected in Finland and Sweden.

The approach of geologic disposal uses both natural and engineered barriers for the isolation of waste. The natural barrier is the host rock at a site and the engineered barrier is the waste package. The isolation is mainly against the infiltrating ground-water. The natural barrier delays the infiltration of groundwater, slows the dissolution of radionuclides, and delays the migration of radionuclides in the geologic medium. The engineered barrier will protect the waste, thus preventing the release of radionuclides into the host rock. If the release does take place, the host rock limits the rate of dissolution and migration of the radionuclides through its chemical and physical characteristics. Through long-term retention of radionuclides in the host rock system, there will be long-time delays for the contaminated groundwater to reach the biosphere, thus minimizing the potential impact of radiation exposure on humans and the ecologic system. As long as both barriers perform well, geologic repositories provide long-term safety to the public.

2.11 Decommissioning of nuclear fuel cycle facilities

Decommissioning is the final step in the life cycle of a nuclear power plant or other fuel cycle facilities. It is the process of taking a plant out of operation and shutting it down permanently. Decommissioning of a nuclear facility requires removal and cleanup of radioactive contamination to allow access by the general public after the termination of the license of the facility.

In the case of decommissioning a nuclear power plant, nuclear fuels and the reactor coolant water are removed first as the largest and mobile source of radio-activity in a nuclear plant. After that the largest majority of the radioactivity in a nuclear power plant remains in the reactor pressure vessel and related components. Depending on the timing of removal of reactor pressure vessel and the related components, and the rest of the contaminated structures in the plant, different options for decommissioning nuclear power plants can be considered. Such options include immediate decontamination and dismantling (sometimes simply called DECON), safe storage (also known as deferred dismantlement, abbreviated as SAFSTOR), and entombment. Under the immediate decontamination and dismantling scenario, all fuel assemblies, radioactive fluids and waste, and other materials having activities above accepted unrestricted activity levels are removed from the site, followed by dismantling the buildings and equipment to permit the removal of regulatory controls. The resulting radioactive wastes are transported and disposed of at appropriate disposal facilities. Dismantlement could take place immediately after cessation of operation. The end state of dismantlement is most likely unrestricted release of the site but some restrictions could apply.

Safe storage (sometimes called mothballing) is "putting the facility in a state of protective storage" as a deferred dismantlement option. Under this option, all spent fuels, the reactor coolants, and existing radioactive waste are first removed from the site. Then the plant is placed in a stable storage condition. Thus reactor vessel and the plant structures remain at the site with security in place. During the storage period (perhaps ranging between 30 and 60 years), the inventory of radioactive materials is reduced through radioactive decay. Decontamination and dismantlement of the reactor vessel and removal of the rest of the plant systems take place after the storage period. Radiation monitoring, environmental surveillance, and appropriate security procedures are established during the storage period.

The third option, entombment, requires the plant to be securely encased in a concrete structure to prevent access and to be maintained with continued surveillance until the radionuclides decay to a level that permits elimination of regulatory controls. Entombment still requires removing all spent fuels, reactor coolants, and any existing radioactive waste, with certain selected components shipped off-site. All the remaining highly radioactive or contaminated components (e.g., reactor vessel and reactor internals) are sealed within a structure integral with the biological shield. The structure should provide integrity over a period of decay of the radioactive materials to reach unrestricted levels. An appropriate and continuing surveillance program should be established but at a reduced level than what is required for SAFSTOR. In reality, uncertainties about the regulatory viability of the option have made the option unattractive. The option has not been exercised in modern nuclear power plant decommissioning.

As the goal of nuclear decommissioning is to terminate the operating license of a nuclear reactor while ensuring public safety after the license termination, the options in decommissioning must include a plan for the end state of the site. If the end state is unrestricted release for public use, it is called "greenfield." If there is any restriction, it is called "brownfield." The "brownfield" option could include some sort of site reuse with restrictions, such as a parking lot, museum, or an industrial facility, or the site could be reused for a new nuclear power plant. Even in the absence of any reuse, the "brownfield" option implies the presence of institutional control after decommissioning for the near future. Under such scenarios, building foundations and other underground structures or piping may be left at the site if the residual contamination levels are very low.

3. Presence or release of radioactive materials in the nuclear fuel cycle

At each step of the nuclear fuel cycle, radioactive materials are present as either naturally occurring or artificial materials. Release of these radioactive materials can occur as part of normal operation of the facilities. Release of a larger amount of radioactive materials is possible under accident conditions.

Impacts from any release of radioactive materials from the nuclear fuel cycle facilities depend on the characteristics and the number of population groups living near the facility. Owing to the site-specific nature of determining such impacts, assessment of the impact of radioactive material release is typically made by assuming a model facility at a representative site with the specified types and the number of population groups in the vicinity. The resulting dose estimate is mostly for the local and regional population. The results are given as the collective effective dose of the affected members of the public, i.e., the sum of all individual effective *doses* over the period of interest.

Radionuclides of concern in the front end of the nuclear fuel cycle are naturally occurring nuclides, with uranium as the most important one. Uranium-238 takes up more than 99% of the mass of the naturally occurring uranium and its radioactive decay chain includes 14 different nuclides. These nuclides are mostly α emitters and preventing intake is one of the most important protection measures. In managing radiation hazard in the front end of the nuclear fuel cycle, protecting the workers who handle these radionuclides in the facilities of the front-end nuclear fuel cycle is an important consideration. Another key nuclide of concern in the front-end nuclear fuel cycle is radon, one of the decay products of uranium. Radon, as a gas, is highly mobile and can be inhaled by the workers or released to the environment away from the fuel cycle facilities as a potential source of radiation exposure to the public.

In the case of a nuclear reactor or the back-end nuclear fuel cycle operations, radionuclides of concern are man-made, i.e., many of them do not exist in nature. The potential impact of these radionuclides could be significant, as they exist in large quantities and are the sources of penetrating radiation such as γ-rays and neutrons. These radionuclides affect the facility workers mostly through direct exposure and the public through airborne and aqueous releases to the environment.

Fig. 3.2 shows each step of the nuclear fuel cycle and the major radionuclides of concern involved as the source of radiation hazard in the fuel cycle. As indicated, uranium and radon are the key nuclides of concern in the front end of the nuclear fuel cycle. In the back end of the fuel cycle, various radionuclides are important in terms of radiation hazards.

In terms of occupation radiation exposure among the workers in the nuclear fuel cycle, the worldwide average data collected between 1990 and 2002 is presented in Table 3.2 (UNSCEAR, 2008). These data are from the 2008 United Nations Scientific Committee on the Effects of Atomic Radiation (UNSCEAR) report. As shown in Table 3.2 the annual effective dose to the monitored workers has steadily been decreasing. The current average effective dose to a worker in a nuclear fuel cycle facility is 1 mSv. This is less than the recommended annual dose limit of 2 mSv to the workers according to the ICRP (1990). This number represents a fourfold reduction from the dose received by the workers in the 1970s (i.e., 4.4 mSv per worker).

The public radiation exposure from the radioactive materials in the nuclear fuel cycle is summarized in Table 3.3 as the worldwide average data. These data are for

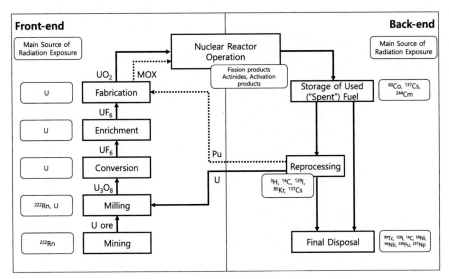

FIGURE 3.2

Radionuclides of major importance released from each stage of the nuclear fuel cycle. *MOX*, mixed oxide.

the period between 1990 and 2002. The annual collective dose to the local and regional population groups due to radionuclide release in effluents of the nuclear fuel cycle has declined from 0.92 person-Sv per GWe in 1990—94 to 0.72 person-Sv per GWe in 1998—2002. The largest contribution was from nuclear reactor operation with 38% contribution followed by uranium mining responsible for 26% of the total.

This result reflects the effort by the industry over the years to reduce the release of radionuclides. While Table 3.3 shows the data only from 1990, the normalized collective effective dose to the public in the 1970s was 12 person-Sv per GWe per year (according to the 2008 UNSCEAR report). This means that the recent dose impact of the nuclear fuel cycle on the public is more than a factor of 16 lower than the dose impact in the 1970s.

3.1 Mining

In underground mines, mining activities and the resulting ore piles release radioactive particles such as ^{238}U and ^{232}Th. Also as the ventilation system is installed to remove dust, blasting fumes, generator exhaust, and radon gas from the mine, the mine exhaust air releases radioactive materials including ^{222}Rn and its decay products. In addition, as most mines are located in deep underground formations, groundwater accumulates in underground mines that is removed through pumping and released to the natural surface drainage system and such water removal operations release radioactive particles.

Table 3.2 Radiation dose to workers in the nuclear fuel cycle (as the world average) (UNSCEAR, 2008).

Stage	Number of monitored workers (thousands)			Average annual collective effective dose (person-Sv)			Average annual collective effective dose per unit energy generated (person-Sv per GWe)			Average annual effective dose to monitored workers (mSv)		
Year	1990–94	1995–99	2000–02	1990–94	1995–99	2000–02	1990–94	1995–99	2000–02	1990–94	1995–99	2000–02
Mining	69	22	12	310	85	22	1.72	0.5	0.1	4.5	3.9	1.9
Milling	6	3	3	20	4	3	0.11	0.03	0.02	3.3	1.6	1.1
Enrichment	13	17	18	1	1	2	0.02	0.02	0.02	0.12	0.1	0.1
Fuel fabrication	21	22	20	22	30	31	0.1	0.1	0.1	1.03	1.4	1.6
Reactor operation	530	448	437	900	779	617	3.9	2.5	2.5	1.4	1.5	1.0
Reprocessing	45	59	76	67	61	68	3.0			1.5	1.1	0.9
Research	120	96	90	90	37	36	1.0	1	1	0.78	0.4	0.4
Total	800	700	660	1400	1000	800	9.8	1	1	1.75	1.4	1.0

Annex B, Table 22.

Table 3.3 Normalized annual collective dose to the public (to the local and regional populations) from the nuclear fuel cycle (UNSCEAR, 2008).

Source	Normalized collective effective dose (person-Sv per GWe)			Major nuclides
	1990 −94	1995 −97	1998 −2002	
Mining	0.19	0.19	0.19	Rn
Milling	0.008	0.008	0.008	Rn (86%), U, th, Ra
Tailings	0.04	0.04	0.04	Rn
Fabrication	0.003	0.003	0.003	U
Reactor operation (airborne effluents)	0.4	0.4	0.22	^{14}C, ^{3}H, noble gas, particulates, iodine
Reactor operation (liquid effluents)	0.05	0.04	0.05	^{3}H, particulates
Reprocessing (airborne effluents)	0.03	0.04	0.028	^{14}C, particulates, ^{3}H, ^{85}Kr
Reprocessing (liquid effluents)	0.10	0.09	0.081	^{14}C, ^{134}Cs, ^{137}Cs, ^{106}Ru, ^{90}Sr, particulates
Transportation	<0.1	<0.1	<0.1	
Total	0.92	0.91	0.72	

Annex B, Table 22.

In surface mining, dust particles containing ^{238}U, ^{232}Th, and ^{222}Rn are released from mining operations, ore piles, and overburden piles as a source of potential public radiation exposure. Both underground and surface mining leave large quantities of solid residue as waste. ISL mining produces solid and liquid wastes but at a much less quantity than underground mining or surface mining. The liquid solution used to leach out uranium from the ore body is collected for treatment for uranium extraction. Release from the storage of liquid solution becomes the source of radon release for public exposure in ISL mining.

As shown in Table 3.2, the average annual effective dose to monitored uranium mine workers in 2002 was 1.9 mSv. This is lower than 4.5 mSv observed in 1990−94. Also the collective dose to the workers from the necessary mining work to produce the amount of natural uranium needed for the generation of 1 GWe of electricity in 1 year was 1.72 person-Sv in 1990−94 but was reduced to 0.1 in 2000−02. Let us call this the normalized collective dose, i.e., the collective dose per GWe. Thus we note that both the individual dose and the collective dose involved in uranium mining have significantly been reduced. This may be in part due to increased use of the ISL method in uranium mining.

The public dose from the release of radioactive materials in uranium mining was 0.19 person-Sv per GWe per year for the period 1998 through 2002. The dose is mostly from the release of ^{222}Rn, as ^{222}Rn takes up the majority of release from

Table 3.4 The estimated amount of release of radionuclides from mining and milling and the resulting collective effective dose (UNSCEAR, 1993).

	Release per unit production (GBq per ton)						
	^{210}Pb	^{210}Po	^{222}Rn	^{226}Ra	^{230}Th	^{234}U	^{238}U
Mining			300				
Milling			13				
	Normalized release (GBq per GWe per year)						
	^{210}Pb	^{210}Po	^{222}Rn	^{226}Ra	^{230}Th	^{234}U	^{238}U
Mining			75,000				
Milling	0.02	0.02	3,000	0.02	0.02	0.4	0.4
Mill tailings (in operation)			3,000				
Mill tailings (closed)			300				

Annex B, Table 19.

mining (Table 3.4) (UNSCEAR, 2000). As shown in Table 3.4, the amount of ^{222}Rn release is three to four orders of magnitude higher than that of ^{238}U or ^{232}Th. Table 3.4 indicates that the amount of ^{222}Rn released per ton of uranium processed is 300 GBq. Also about 75 TBq of ^{222}Rn is expected to be released from the required mining activities to produce the amount of natural uranium needed for the generation of 1 GWe of electricity in 1 year. Overall, to support the production of 1 GWe electricity for 1 year, about 75 TBq of ^{222}Rn is expected to be released from the required mining activities to produce the amount of natural uranium needed.

3.2 Milling

Processing of uranium ore and drying and packaging of yellowcake result in the emission of ^{222}Rn and atmospheric dust containing ^{238}U, ^{230}Th, ^{226}Ra, and ^{210}Pb. This leads to radiation exposure among workers and the public. The average annual effective dose to the monitored workers in uranium milling between 2000 and 2002 was 1.1 mSv (Table 3.2). Also for the same period, the annual normalized collective effective dose to the workers was estimated at 0.02 person-Sv per GWe. Historical trends given in Table 3.2 indicate that the worker dose has been on decline.

In comparison, the annual normalized collective effective dose among the public in the same period was 0.048 person-Sv per GWe. So the total public dose is more than a factor of 2 higher than the total worker dose. This may be because the number of people exposed is large due to the dispersion of radon. We also note that public radiation exposure from uranium milling occurs not only during milling operations but also after the closure of milling operations due to the emissions of ^{222}Rn. Although about 13 GBq of ^{222}Rn is expected to be directly released from 1 ton of natural uranium processing (this is about 3 TBq per GWe) (UNSCEAR, 1993), ^{222}Rn is also released as waste from the pile of tailings accumulated from the milling operation. Annual release of ^{222}Rn from mill tailings accumulated during milling operations and after the facility closure is estimated at 3 and 0.3 TBq, respectively, to support 1 GWe electricity generation. These are summarized in Table 3.4.

3.3 Conversion, enrichment, and fuel fabrication

From the conversion operation of U_3O_8 to UF_6, release of radioactive materials occurs in the form of airborne or liquid effluents. These effluents include uranium and its decay products such as ^{238}U, ^{235}U, ^{234}U, ^{234}Th, and ^{226}Ra. The gaseous effluents are the particulates produced from the processing of feed materials. The liquid effluents are from the process of uranium recovery and from the use of liquid to remove the radioactive particles in the gas streams. The operation of uranium enrichment and fuel fabrication also releases the similar radioactive materials except ^{226}Ra (as radium is separated out from uranium conversion). The levels of radioactivity release from enrichment and fuel fabrication are much less than those in uranium conversion. In fuel fabrication, liquid effluents are produced from the chemical conversion of uranium hexafluoride to uranium dioxide containing chemical impurities.

Table 3.5 shows the details of the expected release. Release of uranium in the order of 100 MBq represents the largest amount of radioactive material release in conversion, enrichment, and fuel fabrication. Solid radioactive waste is produced in conversion, enrichment, and fuel fabrication but it is a very minor issue, as the inventory of radioactive materials is very small.

In uranium enrichment and fuel fabrication, the estimated average annual effective dose to the monitored workers was 0.1 and 1.6 mSv, respectively (see Table 3.2). The estimated normalized annual collective dose to the workers for the generation of 1 GWe of electricity was 0.02 and 0.1 person-Sv in enrichment and fuel fabrication, respectively. The trend shown in Table 3.2 indicates that the worker dose in enrichment and fuel fabrication has been stabilized and is much lower than what is involved in mining or milling.

The normalized collective effective dose to the public from nuclear fuel fabrication in the world between 1990 and 2002 was 0.003 person-Sv per GWe (Table 3.3). The dose is much smaller than that of other fuel cycle facilities.

3.4 Nuclear reactor operation

The level of potential radiation hazard from nuclear reactor operation is the highest in comparison to other stages of the nuclear fuel cycle. A large inventory of radioactive materials is produced from nuclear reactor operation as fission products and actinides. Neutron irradiation of structural and cladding materials, as well as the reactor coolant, also produces various activation products. Many of them are sources of γ-rays and neutrons requiring careful planning and engineering designs for the protection of workers and the public against penetrating radiation.

Table 3.5 Annual release of radionuclides from uranium conversion, uranium enrichment, and fuel fabrication facilities (normalized per GWe in MBq per GWe) (UNSCEAR, 1993).

| Radionuclides | Normalized annual release in MBq per GWe per year | | | | | |
| | Conversion | | Enrichment | | Fuel fabrication | |
	Airborne	Liquid	Airborne	Liquid	Airborne	Liquid
^{226}Ra	0.022	0.11				
^{228}Th	0.4					
^{230}Th	0.022					
^{232}Th						
^{234}Th	130		1.3		0.34	170
^{234}U	130	94	1.3	10	0.34	170
^{235}U	6.1	4.3	0.06	0.5	0.0014	1.4
^{238}U	130	94	1.3	10	0.34	170

Annex B, Table 24.

Fission products are produced when a heavy nuclide undergoes fission. The radionuclides produced include ^{90}Sr, ^{137}Cs, ^{3}H, ^{131}I, ^{129}I, ^{99}Tc, ^{95}Nb, ^{133}Xe, ^{144}Ce, etc. They are important sources of γ-rays and electrons. Although most of the fission product radionuclides are very short-lived, there are 13 fission products with greater than 10 years of half-life. The two most prominent radionuclides of concern for public radiation protection under nuclear accident are ^{131}I ($t_{1/2} = 8.04$ days) and ^{137}Cs ($t_{1/2} = 30.2$ years).

Actinides are the radionuclides produced after uranium and the actinide elements go through neutron capture and subsequent radioactive decays. They include ^{237}Np, ^{239}Pu, ^{240}Pu, ^{241}Pu, ^{242}Pu, ^{241}Am, ^{243}Am, ^{242}Cm, and ^{244}Cm. They form an important part of the radioactive inventory in spent nuclear fuel as important sources of neutrons.

Activation products are the products of neutron capture by reactor structural materials or coolant or fuel impurities. The examples are ^{59}Ni, ^{63}Ni, ^{55}Fe, ^{60}Co from steel activation, ^{93}Zr and ^{95}Zr from cladding activation, and ^{3}H and ^{14}C from the activation of coolant, moderator, or chemicals in the coolant (such as boron and lithium).

In a typical large commercial nuclear power plant, the total inventory of radionuclides present in a nuclear reactor amounts to approximately hundreds of exabecquerel (i.e., 10^{20} Bq). Fission products and actinides take up about 80% and 20% of the total inventory, respectively. The activation products are at less than 0.1% of the inventory. Through fuel failures and corrosion of the structural materials, these radionuclides can become part of the circulating reactor coolant. Presence of these radionuclides in the reactor coolant system causes radiation exposure among workers involved in the operation and maintenance of a nuclear power plant. In particular, worker exposure mostly occurs during scheduled maintenance and/or refueling outages. Most of exposure is from direct exposure from activation products such as ^{60}Co, ^{58}Co, and ^{110m}Ag. In the case of BWRs, ^{16}N is also an important source of worker dose. In HWRs, internal exposure due to ^{3}H could also be significant among the workers. As shown in Table 3.2, the annual worker dose during the operation of nuclear power plants is about 1.4 mSv (as the global average value). The normalized annual collective effective dose among the public between 1995 and 2002 was about 2.5 person-Sv. With 444 nuclear reactors in operation between 1998 and 2002, the average annual generating capacity was about 278 GWe in the world. The corresponding annual collective dose among the public would be 695 person-Sv from the global operation of nuclear power plants.

The amount of radioactive materials released during normal reactor operation as part of plant effluents varies depending on the types of radioactive materials and the reactor design. In terms of atmospheric release, fission product noble gases (e.g., Kr and Xe), activation gases (e.g., ^{14}C), tritium, iodine, and particulates (e.g., ^{88}Rb, ^{89}Rb, ^{139}Ba) take the majority of activity released. For the release through liquid effluents, tritium, fission products (^{90}Sr and ^{137}Cs), and activated corrosion products (e.g., ^{51}Cr, ^{54}Mn, ^{58}Co, ^{60}Co, and ^{95}Zr) take the majority of the inventory released as liquid effluents.

In terms of the details of release to the environment, the annual average data of release for different types of nuclear power plants has been reported by the United Nations (the 2000 UNSCEAR report). The reported data, typically as normalized quantity per GWe of electricity generated, during the period between 1990 and 1994 are summarized in Table 3.6.

For the fission product noble gases (mostly as Xe and some Kr), the normalized amount of release was ~27 TBq for PWR, ~354 TBq for BWR, and 2050 TBq for HWR. The release from GCRs, LWGRs, and FBRs was 1560, 1720, and 380 TBq, respectively. The all-reactor-type average value was 330 TBq. For tritium, the reported normalized release to the atmosphere was 2.3 TBq for PWR, 0.94 TBq for BWR, 650 TBq for HWR, 4.7 TBq for GCR, 26 TBq for LWGR, and 49 TBq for FBR. The all-reactor-type average value was 51 TBq per GWe. In the case of iodine (i.e., ^{131}I), the normalized release was 0.33 GBq for PWR, 0.81 GBq for BWR, 0.35 GBq for HWR, 1.4 GBq for GCR, and 6.8 GBq for LWGR. The all-reactor-type average was 0.81 GBq. For airborne release of particulates, PWRs, BWRs, HWRs, GCRs, LWGRs, and FBRs each emitted 0.18, 178, 0.051, 0.30, 14, and 12 GBq per GWe. The all-reactor-type average was 40 GBq per GWe. While the reactor-type-specific values were not available, the annual average value of ^{14}C release was 0.44 TBq per GWe.

In terms of liquid effluents the key radionuclide released was ^{3}H. PWRs, BWRs, HWRs, GCRs, and FBRs each emitted 22, 0.94, 490, 220, and 1.8 TBq of ^{3}H per GWe, respectively. The all-reactor-type average value was 51 GBq per GWe. The release of all other radionuclides as the combined normalized quantity in the liquid effluents was 19 GBq for PWR, 43 GBq for BWR, 130 GBq for HWR, 510 GBq for GCR, 4.8 GBq for LWGR, and 49 GBq for FBR. The all-reactor-type average value was 48 GBq per GWe.

According to the 2000 UNSCEAR report, the atmospheric release was responsible for 89% of the dose, with the remaining 11% from the aquatic release. The largest contributor to the dose was ^{14}C (28%), followed by noble gas (26%) and particulates (18%) all from atmospheric release. Tritium contributed to the 17% and 7.2% of the dose from atmospheric and aquatic releases, respectively. Other nuclides contributed to 3.7% of the dose from aquatic release. The contribution from ^{131}I through atmospheric release was less than 0.05%. These are summarized in Table 3.5.

Various types of solid radioactive waste are also produced in nuclear power plant as radioactive species are distributed throughout the nuclear power plant. These wastes are called low-level waste in the United States or low- and intermediate-level waste outside the United States. A large portion of these wastes is generated as byproducts from the treatment of radioactive liquid and gaseous effluents in nuclear power plants. These wastes include spent ion-exchange resins, discarded filter material, or evaporator residue. Other solid radioactive wastes include various types of activated reactor hardware, discarded equipment and tools, and contaminated trash (dry active waste) from day-to-day operation, maintenance, and modification activities. The radioactivity in these wastes is mostly as activation products (e.g.,

Table 3.6 Annual release of radionuclides from nuclear reactors and the associated collective effective dose (based on the data reported for the period 1990–94) (UNSCEAR, 2000).

| | Normalized release per GWe of electricity generation (TBq per GWe per year) | | | | | | |
| | Airborne release | | | | | Liquid release | |
	Noble gas	^{3}H	^{14}C	^{131}I	Particulates	^{3}H	Other
PWR	27	2.3	0.22	0.0003	0.0002	22	0.019
BWR	350	0.94	0.51	0.0008	0.18	0.94	0.043
GCR	1600	4.7	1.4	0.0014	0.0003	220	0.51
HWR	2100	650	1.6	0.0004	0.00005	490	0.13
LWGR	1700	26	1.3	0.007	0.014	11	0.005
FBR	380	49	0.12	0.0003	0.012	1.8	0.049
Total	330	36	0.44	0.0007	0.040	48	0.047

Total combined release = 415 (TBq)

| | Normalized collective effective dose from radionuclides released per year (person-Sv per GWe) | | | | | | |
| | Airborne release | | | | | Liquid release | |
	Noble gas	^{3}H	^{14}C	^{131}I	Particulates	^{3}H	Other
PWR	0.003	0.005	0.059	0.0001	0.0004	0.014	0.006
BWR	0.15	0.002	0.14	0.0002	0.36	0.0006	0.014
GCR	1.44	0.010	0.38	0.0004	0.0006	0.14	0.17
HWR	0.23	1.4	0.43	0.0001	0.0001	0.32	0.043
LWGR	0.19	0.05	0.35	0.002	0.028	0.007	0.002
FBR	0.042	0.10	0.032	0.00009	0.024	0.0012	0.016
Weighted average	0.11	0.075	0.12	0.0002	0.080	0.031	0.016

Total combined collective dose = 0.43 (person-Sv)

Annex C, Table 37, 39.
BWR, *boiling water reactor;* FBR, *fast breeder reactor;* GCR, *gas cooled reactor;* HWR, *heavy water reactor;* LWGR, *light water graphite reactor;* PWR, *pressurized water reactor.*

^{60}Co, ^{63}Ni, ^{55}Fe) and fission products (e.g., ^{137}Cs and ^{90}Sr). These radioactive wastes are normally stored on-site and periodically shipped to authorized storage facilities for low-level waste. These wastes are stabilized or solidified and disposed of in a disposal facility for low-level waste or low- and intermediate-level wastes by using metallic or high-density polymer containers. The annual dose impact to the public from the disposal of the solid radioactive wastes from nuclear power operation is estimated at around 0.5 person-Sv per GWe (UNSCEAR, 2000).

Under accident conditions, release of radioactive materials to the environment can take place in a much larger quantity. As mentioned, key radionuclides of concern in this case are ^{131}I ($t_{1/2} = 8.04$ days) and ^{137}Cs ($t_{1/2} = 30.2$ years). In the case of three major nuclear accidents that took place in the past, the amount of release is estimated as follows:

> The Three Mile Island (TMI) accident: 550 GBq of ^{131}I released to the atmosphere (UNSCEAR, 2008, Volume II, p. 6).
> The Chernobyl accident: 1760 PBq of ^{131}I and 85 PBq of ^{137}Cs released to the atmosphere (UNSCEAR, 2000, Annex J. p. 457).
> The Fukushima accident: 100−500 PBq of ^{131}I and 6−20 PBq of ^{137}Cs released to the atmosphere (UNSCEAR, 2013, Chapter III, p. 6).

The dose impact of these accidents is estimated at 20 person-Sv (TMI), $0.6−2.4 \times 10^6$ person-Sv (Chernobyl), and up to 48,000 person-Sv (Fukushima) (UNSCEAR, 2013, Annex A, Table 8). Note that the collective dose among the public is about 700 person-Sv from global operation of nuclear power plants in 1 year. Thus the release from the Chernobyl accident was equivalent to the combined release from a global nuclear power plant operation over up to 3430 years. The Fukushima accident release was equivalent to the combined release from 69 years of global power operation. From this perspective, the TMI can be considered a minor event.

3.5 Reprocessing

By the time the spent fuel reaches the stage of reprocessing, most of the shorter lived fission products (i.e., ^{131}I, ^{133}Xe, ^{140}Ba-La) in the spent fuel will have decayed away, but longer lived fission products and activation products remain in the spent fuel.

As part of the so-called head-end process of reprocessing, fuel cladding is opened by mechanical shearing and UO_2 fuel is chopped into short segments. During this process, most of the gaseous radionuclides inside the cladding are released as off-gases. The sheared short segments of UO_2 fuel are then exposed to air or oxygen at a high temperature. This results in the conversion/oxidation of UO_2 to mostly U_3O_8. In this process the fuel structure is broken up releasing gaseous or volatile radionuclide products. The off-gases from the fuel dissolution process also contain gaseous radionuclides. These radionuclides include ^{3}H, ^{14}C, ^{85}Kr, ^{90}Sr, ^{106}Ru, ^{129}I, ^{134}Cs, and ^{137}Cs.

Although these off-gases are treated to minimize radiation release from the facilities, a small fraction of radionuclides is released to the atmosphere. Krypton-85 represents the largest inventory of radioactivity released. The normalized amount of annual release of ^{85}Kr is 6330 TBq per GWe. This is followed by ^{3}H with 24 TBq per GWe. In terms of liquid effluents the largest release is that of ^{3}H with 270 TBq annual release per GWe. Other major nuclides released in liquid effluents include ^{90}Sr, ^{106}Ru, ^{129}I, ^{137}Cs, and ^{14}C. The details of the release are summarized in Table 3.7 (UNSCEAR, 2000). Solid wastes that also arise from reprocessing include fuel-cladding hulls, particulate filters, discarded equipment tools, and contaminated trash including most of these radionuclides (except ^{85}Kr and ^{106}Ru).

The average value of annual collective effective dose among the workers was estimated at 68 person-Sv from global reprocessing operations between 2000 and 2002. This value has been somewhat steady since 1990. The average value of the annual occupational dose due to reprocessing operation is 0.9 mSv per person between 2000 and 2002. This per person dose has been in steady decline since the 1970s from 7.1 mSv in 1995—99.

In terms of public impact, the normalized annual collective effective dose between 1990 and 2007 was estimated at 0.13 person-Sv per GWe from both liquid and airborne releases. The number slightly decreased to 0.11 person-Sv per GWe between 1998 and 2002. The liquid pathway and the airborne pathway each contributed to 69%—74% and 23%—31% of the dose, respectively. The largest contributor to the dose to the public was ^{14}C.

4. Concern with spent nuclear fuel

As described earlier, the fresh nuclear fuel for most nuclear reactors is made of uranium in the form of oxide (UO_2). But as the uranium in nuclear fuel is irradiated in a nuclear reactor, it undergoes nuclear transformation that results in drastic changes in the composition, including a diverse mixture. This mixture includes a variety of fission products and actinides. Actinides refer to elements that range in atomic numbers (Z) from 89 to 103.

Compared with fresh fuel, the content of ^{235}U in spent fuel is reduced from about 3%—4% (weight percent) to less than 1% in the fuel. The ^{238}U content is reduced from about 96%—97% to about 93%—94%. Fission products are produced, as a replacement of the lost mass, in the amount of about 3%—4%. Plutonium and other actinides collectively take up about 1% of mass in spent fuel. These actinides are also called minor actinides. The detailed composition of spent fuel varies depending on how long the fuel is irradiated in the reactor (i.e., as a function of burnup) or how long it has been cooled since discharge from the reactor.

Important radionuclides in spent fuel as radiation source typically include the following:

Table 3.7 Release of radionuclides from reprocessing and the resulting collective dose (during the period 1990–94, based on the total cumulative fuel reprocessed amount equivalent to 130 GWe-year) (UNSCEAR, 2000).

	Airborne release						Liquid release					
	^{3}H	^{14}C	^{85}Kr	^{129}I	^{131}I	^{137}Cs	^{3}H	^{14}C	^{90}Sr	^{106}Ru	^{129}I	^{137}Cs
Release per GWe of electricity generation for 1 year (TBq per GWe per year)	24	0.4	6300	0.001	9E-5	8E-5	270	0.8	2.0	2.1	0.03	1.0
Collective effective dose per unit release (person-Sv per TBq)	0.0021	0.27	7.4E-6	44	0.3	7.4	1.4E-6	1.0	4.7E-3	3.3E-3	0.099	0.098
Total collective effective dose in 5 years among regional population (person-Sv)	6.6	13	6.1	8.4	0.003	0.08	0.05	98	1.2	0.9	0.4	12

Annex C, Table 41.

- γ Source:
    - ^{60}Co, ^{90}Sr, ^{90}Y, ^{134}Cs, ^{137}Cs, ^{137m}Ba, ^{144}Pr, ^{154}Eu, ^{244}Cm, etc.
- Neutron source:
    - ^{244}Cm (dominant source), ^{241}Am, ^{246}Cm, ^{240}Pu, ^{242}Pu, ^{238}Pu, etc.

In terms of radiation hazard presented by spent fuel, up to several hundred grays per hour of absorbed dose is possible, in the case of an unshielded spent fuel. Regardless of the burnup level, ^{60}Co, ^{90}Y (the decay product of ^{90}Sr), and ^{137m}Ba (the decay product of ^{137}Cs) are important sources of γ radiation in spent fuel. With much less importance, ^{154}Eu and ^{134}Cs follow them as γ-ray source. The importance of actinides as the source of neutrons depends on the level of burnup. At low burnup, the actinides are of minor importance as the fuel did not have enough time to build them. At high burnup, enough inventory of ^{244}Cm and ^{246}Cm is produced, which becomes an important contributor to neutron dose.

Presence of γ-rays and neutrons in spent fuel demands installation of proper shielding. Shielding should provide the necessary level of safety to protect the workers and the public during handling, shipping, and storage of spent fuel. For shielding against γ-rays, lead or depleted uranium is often used. Depleted uranium is a byproduct from enriching natural *uranium* and has a residual ^{235}U content of 0.2%−0.3%, with ^{238}U comprising the remainder. The material for neutron shielding is typically water or solid organics implemented in the form of plastics or resins. Use of water for cask shielding is rarely exercised due to possible loss under impact. Use of materials for inelastic scattering of neutrons at energies of several hundred kiloelectronvolts or above is often necessary to effectively reduce neutron energy to the thermal region where the probability of neutron absorption is much higher. For such purposes, iron is also used as typical γ shield material in support of neutron shielding.

Exposure to radiation among the public occurs during the transportation of spent fuel. Therefore use of special packaging is needed for the transport of spent fuel is known. Such special packaging is called a shipping cask. Shipping casks are massive and heavy reusable vessels designed to provide physical containment, radiation shielding, heat removal, criticality protection, and theft protection.

As transport of spent fuel is typically done by rail or truck, special purpose casks are developed and used, with their design and size being different depending on the mode of transportation. The casks consist of a large cylindric steel vessel, shielding layers, fuel baskets, a bolt-down lid, and impact limiters. Fuel assemblies are placed inside fuel support baskets within the inner cavity which are designed to accept different types of assemblies. Such spent fuel loading operations always take place under water for the purpose of safety. The baskets provide mechanical support structure and criticality control by including boron inside the structural material. To load fuel into the cask, the cask containing the fuel basket is placed into the spent fuel pool, and spent fuel assemblies are lowered into the basket.

The fuel baskets are surrounded by two layers of shielding inside the vessel. The first shield layer that surrounds the fuel cavity is the γ-ray shield, made of lead, depleted uranium, or steel. The second layer is the neutron shield (sometimes with the added feature of cooling) and consists usually of boron in a mixture of water and ethylene glycol jacket or a specially prepared polyester resin compound. The two layers are sandwiched between steel shells. The outer side of the cask is solid steel. The casks are closed by a massive bolt-down lid. Emplacement of shields aims at reducing the dose level to be less than 20 mSv/hour at the surface of the cask and 0.1 mSv/hour at 2 m from the cask. In case of an accident, the dose level at 1 m from the surface of the cask should be less than 10 mSv/hour.

To limit the impact of collision accidents through attenuation, impact limiters are placed at each end of the vessel. As an example of shield for spent fuels against γ-rays and neutrons in a shipping cask to transport 4 assemblies of PWR spent fuel, a combination of lead (as γ shielding) and solid resins (as shielding against neutrons) was used outside the fuel basket, with the thickness of lead and resins being 15.5 and 13.66 cm, respectively (Kang et al., 1988).

Integrity of spent fuel casks must be demonstrated even under severe accidents. For such purposes, casks are designed to withstand a sequence of hypothetical tests without releasing more than a specified small amount of radioactive material. Such tests include sequential exposure of the cask to impact by drops, fire (800°C for 30 minutes), and water immersion (for 8 hours). The impact tests include a free 9-m drop of the cask on to a flat, horizontal, unyielding surface and a 1-m drop onto a 15-cm-diameter steel bar at least 8 inches long with the bar striking the cask at its most vulnerable spot. Manufacture, testing, use, and maintenance of casks must be carefully conducted to ensure their integrity. The condition of seals and cask components must be checked regularly as part of maintenance requirements.

Radiation dose from truck or rail cask during incident-free shipment was estimated to be less than 0.1 mSv/hour at 1 m from the surface of truck cask or rail cask for an average of PWR and BWR spent fuels (Sprung, 2000). The values vary mainly based on the period of spent fuel cooling. The dose rate is lower with rail casks owing to heavier shielding. If the spent fuel has been cooled for longer than 10 years, the dose rate is less than 0.07 mSv/hour for all cask types. Actual dose to the public or the members of the shipment crew will depend on the duration of exposure, the distance from the spent fuel, and placement of shielding. According to the 2008 UNSCEAR report, the maximum annual dose to a person among the members of the public due to the transport of spent fuel is estimated at less than 4 μSv for trucks, less than 6 μSv for rail, and less than 1 μSv for sea-based shipment (UNSCEAR, 2008, Annex B, Table 28).

In the case of an accident during transportation of spent fuel, the dose to the population was estimated at 8.0×10^{-9} (person-Sv) for truck accidents and 9.4×10^{-8} (person-Sv) for train accidents (UNSCEAR, 2008). The values are based on considering all exposure pathways including inhalation, ingestion, and direct exposure and are the average of the total combined radiation risk among all exposed people during the accident. The results indicate that the consequence of spent fuel transportation accident is not significant because of the very low probability of cask failure from major accidents.

As to the level of radiation hazards from the final disposal of spent fuel or HLW, the responsible national authorities typically designate a fraction of the 1 mSv per year as a limit of dose to the public for nuclear waste disposal facilities. The resulting dose limit for the public from nuclear waste disposal ranges from 0.05 to 0.3 mSv per year. This indicates that the level of safety imposed in nuclear waste disposal is one order of magnitude lower than the natural background radiation exposure level.

---

5. Radiation concern in nuclear power plant decommissioning

Decommissioning of nuclear power plants normally takes place within a period of about 100 years after the shutdown of a nuclear reactor. Major radionuclides of concern as a source of radiation exposure to workers are ^{60}Co, ^{58}Co, ^{63}Ni, ^{59}Ni, and ^{94}Nb.

Cobalt-60 is the most important nuclide of concern in decommissioning. It is a strong γ emitter with a half-life of 5.2 years. On average, ^{60}Co is known to be responsible for over 80% of the dose to workers during the operation of a plant. Along with ^{60}Co, ^{58}Co is also important at reactor shutdown and for about 4 years, as both ^{60}Co and ^{58}Co mainly control the radiation levels in the reactor. ^{60}Co continues to be dominant in controlling the radiation levels up to 30 years. If decontamination and decommissioning takes place right away after the plant shutdown, the worker dose during the decommissioning operations will be high due to ^{60}Co. Waiting for about 40 years for decommissioning will have the benefit of most of ^{60}Co decayed away as the main rationale for the SAFSTOR option. Therefore the choice between immediate or delayed decommissioning involves cost trade-offs between costs of storage and surveillance versus the higher costs of handling the higher radiation levels associated with earlier decommissioning. After that, ^{63}Ni, a β (66 keV) emitter takes a dominant place in the plant's radiation levels up to several hundred years. At longer periods beyond hundreds of years, ^{59}Ni (γ- and X-ray emitter) and ^{94}Nb (γ and β emitter) each with half-life of 8×10^4 and 2×10^4 years, respectively, become an important source of radiation in the plant components. If a significant inventory of ^{59}Ni and ^{94}Nb inventory exists in a reactor facility from a long operating history, the required time to achieve unrestricted release would be too long and exercising the option of entombment becomes unrealistic.

Decommissioning of a nuclear power plant generates a wide variety of radioactive waste. Aside from spent fuels, reactor pressure vessel, steam generators, and other large equipment such as reactor coolant pumps and pressurizers are all radioactive waste. The decontamination and dismantling process and the site remediation activities will also produce a large volume of radioactive waste. These wastes include contaminated concrete, pipes, ducts, miscellaneous steels, and secondary radioactive waste from decontamination and effluent treatment. Also a large volume of slightly contaminated soil is also expected to be generated from the site remediation activities. The resulting waste management activities follow similar processes to radioactive waste management activities during the operation of the reactor. But minimization of radioactive waste generation will play a very important role in nuclear decommissioning.

According to the experiences in the United States with the decommissioning of 13 nuclear power plants between 1995 and 2002, the annual collective effective dose among the workers was about 4 person-Sv with an individual average annual dose of 1.96 mSv (UNSCEAR, 2008, Volume 1, Table 68).

6. Conclusions

Radiation is a carcinogen. Human exposure to radiation must be carefully controlled. As all stages of the nuclear fuel cycle are related to the release of radioactive materials, proper regulatory control must be in place to minimize such release.

As shown in Table 3.3, the largest impact from the nuclear fuel cycle is from nuclear reactor operation followed by reprocessing operation. The main radionuclides contributing to the local and regional population from nuclear power operation are ^{14}C and tritium. In the case of a reprocessing operation, fission products such as cesium (^{134}Cs and ^{137}Cs) and ruthenium are found to be the main dose contributor. If we extend the impact of nuclear fuel cycle operations to the global populations over long-term future, including the impact from the disposal of solid radioactive waste, the estimated normalized collective dose is about 50 person-Sv over 10,000 years.

To put the radiologic impact from the nuclear fuel cycle into perspective, dose from various sources of ionizing radiation needs to be compared. This comparison is shown in Table 3.8 based on the 2008 report by the United Nations (UNSCEAR, 2008). As shown in Table 3.8, human radiation exposure in terms of annual average dose is mostly from natural sources, contributing 80% of the total radiation dose. The largest source of radiation dose to humans is radon gas. The artificial sources contributed 20% of the total. The average dose to an individual among the public is about 0.2 µSv per year. In comparison to other sources of radiation exposure, the average individual radiation dose from the nuclear fuel cycle is less than 0.007% of the total. If the dose from the Chernobyl accident is added to the dose from the nuclear fuel cycle, the contribution becomes 0.073%.

The nuclear fuel cycle does add to the radiation dose received by the public. However, it can be said that contributions from the nuclear fuel cycle to human radiation exposure are not significant in comparison to the radiation dose received by the public due to the presence of natural background radiation.

Table 3.8 Annual dose to humans from various sources of ionizing radiation as world average (UNSCEAR, 2008).

	Source	Annual dose as world average (mSv/year)	Typical range of individual doses (mSv/year)
Natural	Radon gas (inhalation)	1.26	0.2–10
	External terrestrial	0.48	0.3–1
	Ingestion of food	0.29	0.2–1
	Cosmic radiation	0.39	0.3–1
	Total natural	2.4	1–13
Artificial	Medical diagnosis	0.6	0–2.0
	Atmospheric nuclear testing	0.005	Declining from 0.11 mSv/year in 1963
	Occupational exposure	0.005	0–20 (highest dose is from radon among miners)
	Chernobyl accident	0.002	Decreasing from a maximum of 0.04 mSv/year in 1986
	Nuclear fuel cycle (public exposure)	0.0002	Could go up to 0.02 mSv/year for some critical groups
	Total artificial	0.6	0–several tens

Annex B, Table 12.

References

ICRP, 1990. Recommendations of the international commission on radiological protection. ICRP publication 60. Annals of the ICRP 21 (1–3), 72. International Commission on Radiological Protection (1991).

ICRP, 2007. The 2007 Recommendations of the International Commission on Radiological Protection, ICRP Report 103. The International Commission on Radiological Protection.

Kang, H.Y., et al., 1988. Shielding design of shipping cask for 4 PWR spent fuel assemblies. Journal of Korean Nuclear Society 20 (1), 65–70.

Sprung, J., 2000. "Reexamination of Spent Fuel Shipment Risk Estimates," NUREG/CR-6672, vol. 1. Sandia National Laboratory, pp. 3–47.

UNSCEAR, Sources and Effects of Ionizing Radiation, 1993. United Nations Scientific Committee on the Effects of Atomic Radiation, UNSCEAR 1993 Report to the General Assembly, With Scientific Annexes.

UNSCEAR, Sources and Effects of Ionizing Radiation, 2000. Volume 1: Sources, United Nations Scientific Committee on the Effects of Atomic Radiation, 2000 Report to the General Assembly, With Scientific Annexes.

UNSCEAR, Sources and Effects of Ionizing Radiation, 2008. Volume 1: Sources, United Nations Scientific Committee on the Effects of Atomic Radiation, 2008 Report to the General Assembly, With Scientific Annexes A & B.

UNSCEAR, Sources, Effects, and Risk of Ionizing Radiation, 2013. UNSCEAR 2013 Report, Volume I, Report to the General Assembly, Scientific Annex A: Levels and Effects of Radiation Exposure Due to the Nuclear Accident after the 2011 Great East-Japan Earthquake and Tsunami.

Health effects of exposure to ionizing radiation

4

Wilner Martinez-López, MD, PhD [1,2], **Manoor Prakash Hande, PhD, MPH** [3]

[1]*Epigenetics and Genomic Instability Laboratory, Instituto de Investigaciones Biológicas Clemente Estable, Montevideo, Uruguay;* [2]*Associate Unit on Genomic Stability, Faculty of Medicine, University of the Republic (UdelaR), Montevideo, Uruguay;* [3]*Department of Physiology, Yong Loo Lin School of Medicine and Tembusu College, National University of Singapore, Singapore*

1. Introduction

In 1895, Wilhelm Roentgen has discovered the X-rays (Glasser, 1995), and in 1896, Becquerel observed that uranium emitted radiation, termed later on radioactivity by Marie Curie (Radvanyi and Bordry, 1984, 1988). Although, the discovery of adverse effects produced by ionizing radiation occurred soon after the discovery of X-rays in 1895, the health effects of ionizing radiation have been properly determined by long-term, large-scale epidemiological studies, related to Japanese atomic bomb survivors of Hiroshima and Nagasaki (the so-called Life Span Study [LSS]). In this respect, the LSS was employed for the protection system of the International Commission on Radiological Protection (ICRP) (2007).

In general terms, it was clearly demonstrated that atom bomb survivors have a radiation-related excess risk of cancer throughout their whole life, and radiation-exposed children possess a higher risk of developing cancer. On the other hand, it was determined that at high absorbed radiation doses increase the risk of cardiovascular disease as well as other noncancer diseases. However, there is a certain concern in occupational and medical exposures to ionizing radiation, indicating that a dose-response relation needs to be established for effects of radiation at the low-dose range (Kamiya et al., 2015; Little, 2003a,b), since the use of ionizing radiation has increased significantly during the past 2 decades or so, especially in diagnostic and therapeutic applications for nontransmissible chronic diseases such as cardiovascular and cancer, which represent the highest incidence of mortality all over the world (Bray et al., 2018). Historically, humankind has witnessed few nuclear accidents/incidents as well: the Hiroshima and Nagasaki bombing (1945), Chernobyl (1986), Goiânia (1987), and Fukushima (2011) to name a few. Moreover, we all live in an environment where radiation-emitting devices are abundant, and it appears that radiation exposure is practically unavoidable. Thus, obviously, it is important to understand the biological effects of various types of radiation.

Advanced Security and Safeguarding in the Nuclear Power Industry. https://doi.org/10.1016/B978-0-12-818256-7.00004-0

2. Basic concepts of ionizing radiation

Briefly, radiation consists on propagation of energy in the form of either electromagnetic waves (or photons) or subatomic particles traveling in the air. Ionizing radiation corresponds to a radiation with sufficient energy to traverse matter (i.e., biological material) producing ions and free radicals. Therefore, energy deposition through the ionization of atoms and molecules can cause chemical changes. In this respect, it can be defined as linear energy transfer (LET) as the energy per unit length transferred to material when an ionizing wave/particle travels through it. It is measured in keV/μm, and the value varies with different types of radiation from a few keV/μm (such as X-rays or gamma rays with low linear energy transfer) to more than 1000 keV for heavy ions (α- or β-particles as well as neutrons with high linear energy transfer). The quantity of energy deposited by ionizing radiation in a defined mass of material (i.e., biological tissues) is termed the absorbed dose and is measured in J/kg, and the unit name is Gray (Gy). Besides, for radiation protection purposes, the ICRP (2007) has defined two dose concepts: the equivalent dose and the effective dose; in both cases, the SI unit (*Système international (d'unités)/International System of Units*) is the Sievert (Sv). These doses are used in the radiation protection system to account for the extent of microscopic damage in different organs and tissues. "The equivalent dose is the sum of the absorbed doses to an organ or tissue, each multiplied by the radiation weighting factor for the type of radiation, showing the damage-weighted risk of stochastic effects resulting from low-level exposure to that radiation. The value for the radiation weighting factor is defined as 1 for γ-rays and other sparsely ionizing radiation (low linear energy transfer), whereas for densely ionizing radiation with high biological effectiveness (high linear energy transfer), the factor is more than 1, and for α-particles, the weighting factor is 20" (ICRP, 2007).

The penetration of X-rays or γ-rays into biological tissue is much deeper than that of α- or β-particles (Furusawa, 2014; Pouget and Mather, 2001). Therefore, the energy deposition and the subsequent damage induced by X-rays or γ-rays are spread throughout the tissue, whereas α- or β-particles as well as neutron deposit more energy along its track causing local damage. Generally, radiation effects can be grouped into two categories: deterministic effects and stochastic effects. Deterministic effects are based on cell killing and characterized by a threshold dose below which no clinical effect can be observed. Long-term and low-level (chronic) exposure to radiation usually produces stochastic effects (ICRP, 2007). Radiation-induced injuries are dose dependent, and the extent of injury increases with dose. Usually, risks from exposures to low doses of radiation (below 0.1 Gy) can be estimated by extrapolating from data obtained after exposure to higher doses of radiation following the linear model without a threshold or the linear no threshold (LNT) model (Hamada and Fujimichi, 2014). In this chapter, biological effects of ionizing radiation are reviewed.

3. Biological effects of ionizing radiation

Radiation can ionize any atom or molecule inside the cell and produce active free radicals, which can produce chemical changes on any cell component. These changes may either damage essential cellular functions or alter the genetic information, as well as can potentially kill the cell or prevent it from reproducing. Main target for radiation is genetic information of the stem cells. Biological effects of radiation produce genetic damage through a direct or indirect mechanism (through the effect of free radicals), producing changes in the DNA molecule structure. In the direct action of radiation, the ionization track hits the DNA molecule directly and disrupts the molecular structure producing the breakage of one (single-strand breaks or SSBs) or both (double-strand breaks or DSBs) strands of DNA, which eventually could lead to cell death. However, cells that survive radiation damage may later induce cancer and/or other abnormalities. Induction of cancer and associated abnormalities becomes predominant with high-LET (densely ionizing) radiations such as α-particles and neutrons, as well as with high-radiation doses (more than 1 Gy). In the indirect action, radiation hits the water as well as other organic molecules, producing free radicals and hydrogen peroxide, which can induce structural DNA damage. Since water represents the 70% of cell composition, most of the ionizing radiation—induced damage comes from the indirect mechanism. Another interesting issue is that biological and physiological alterations produced by the effect of ionizing radiation could be developed many years later.

Radiation-induced DNA damage can be expressed as mutation occurring in genes or chromosomes of stem cells, which can alter the genetic information that passes from a cell to its progeny. As an inherent mechanism, most DNA damage and mutations are efficiently repaired or corrected by a variety of DNA repair mechanisms; however, these repair processes are error prone. Although most of the DNA damage is repaired, some damage remains or is incorrectly repaired, having consequences on the well-being of the cell and its progeny. Eventually, such persistent damages can have effect on the tissues and organs. On the basis of the DNA breakage (SSB and DSB) produced by ionizing radiation, in 1973, the linear quadratic (LQ) equation was formulated, based on the idea that low doses of ionizing radiation should essentially cause SSBs, easily repairable, whereas high doses would cause the breaks of both strands of the double helix of DNA, potentially lethal to the cell. Besides, not only genetic but also epigenetic changes could be in the base of the evolution of these alterations (Koturbash et al., 2008). There are excellent reviews on this topic available in the literature, to name a few (Kamiya et al., 2015; Little, 2003a,b; Upton et al., 1992; Vaiserman et al., 2018); here, we try to provide a general overview on the health impacts of ionizing radiation.

4. Acute effects of ionizing radiation

Medical practitioners started to regulate and determine the radiation limit exposures only after the World War I, although harmful effects of higher doses of ionizing radiation were identified many years earlier. The International X-ray and Radium Protection Committee established in 1921 (presently—International Commission on Radiological Protection [ICRP]) proposed in 1924 a permissible dose rate for radiation workers up to what today correspond to 500 mSv/year (Mutscheller, 1950) while "the current exposure limits are much lower. For example, ICRP sets the occupational exposure limit at 20 mSv/year (25 times lower), and the limit for the public at 1 mSv/year (500 times lower)" (ICRP, 2007; Vaiserman et al., 2018).

Significant damages to tissues or human health could be produced by exposures in the order of 1−2 Gy or more, and deterministic and stochastic effects can be observed immediately. The massive exposure to ionizing radiation on proliferating tissues (such as bone marrow, blood, and epithelial cells) could cause death of millions of cells, and the final effect is proportional to the extent of damage and duration of the exposure, producing bone marrow aplasia, bleeding, coma, and finally death within minutes or hours, whereas anemia, aging, and diarrhea could be induced in more diluted massive exposures. Stochastic effects, on the contrary, depend on the total dose of ionizing radiation−induced DNA damage on genes controlling cell cycle, programmed cell death (apoptosis), DNA repair process, protooncogenes, or tumor suppressor genes, leading to the development of cancer and other disorders (Nussbaum and Kohnlein, 1994, 1995).

4.1 Main overexposed cases to ionizing radiation

Atomic bombs released in two cities of Japan in 1945 as well as one of the largest nuclear disasters in the human history in Chernobyl (Ukraine) nuclear power plant, plus the most recent accident in another nuclear power plant from Fukushima Daiichi in 2011, are clear examples of the negative side of the scientific progress. Most of the epidemiological data concerning the effects of radiation exposures were obtained through analysis of cohorts from Hiroshima and Nagasaki bombing survivors, as well as from the fallout of radionuclides after the Chernobyl accident (Kamiya et al., 2015).

4.1.1 Atomic bombs at Hiroshima and Nagasaki

In August 1945, two atomic bombs were dropped on the Japanese cities of Hiroshima and Nagasaki. These bombs contained uranium-235 and plutonium-239 that were dropped in Hiroshima and Nagasaki, respectively. As per local authorities in Japan, approximately 140,000 people died in Hiroshima and 74,000 died in Nagasaki within short period of time after the incident (Kamiya et al., 2015). Subsequently, the LSS was established in 1950 through which a systematic epidemiological study of the Japanese atomic bomb survivors was imitated (Ozasa et al., 2018). These studied helped immensely on our understanding of radiation effects on human health. The experience gained from the LSS and from medical care of atomic bomb survivors made great contribution in the creation of the Fukushima Health Management Survey and the medical care of local residents after the 2011 Fukushima Daiichi nuclear power plant accident (Yasumura et al., 2012).

High doses of radiation produced acute deterministic effects, including death from severe gastrointestinal and bone marrow damage as well as injuries from the blast and heat from the bombs. On the other hand, tissues from thousands of victims, irreversibly damaged by particles and neutrons, developed a great number of diseases during the following years or decades both among survivors and their offspring. In case of cancer, there was a significant increase in leukemia cases (excluding chronic lymphatic leukemia), during years immediately following the bombing (Folley et al., 1952). Besides, epidemiologists detected a significant increase of solid tumors in various tissues and organs some years later, especially thyroid, breast as well as lung cancer (Land et al., 1993; Pierce et al., 2005; Ron, 2003). The atom bomb survivor cohort has been, up to now, the best population for estimating cancer risk resulting from radiation exposure based on the data accumulated in the LSS. However, this large cohort of round 100,000 survivors was not adequate to offer statistically significant and conclusive results for impact of low-dose radiation exposure (Nussbaum and Kohnlein, 1994; Barcellos-Hoff, 2008).

4.1.2 Chernobyl accident

The nuclear power plant accident at Chernobyl released a huge amount of radionuclides (iodine, mainly ^{131}I and cesium, mainly ^{134}Cs and ^{137}Cs), into the atmosphere. The amount of people exposed to substantial levels of fallout was probably over 20 million in addition to the hundreds of workers who operated for 2 weeks to cover the molten core from above who were directly exposed to total body irradiation. Food, especially milk, represented the most important source of contamination. Since internal and external exposures were mixed and information was not disclosed immediately after exposure, dose assessment was very difficult (Drozdovitch et al., 2013). In this respect, Chernobyl differs substantially from Hiroshima and Nagasaki nuclear accident, since in the latter one, overexposure was largely due to external radiation resulting from the explosion of the bombs. However, the effects of psychological and social factors were very important in both nuclear accidents (Bromet et al., 2011). It is estimated that thyroid gland received an average ^{131}I dose of 650 mGy in Ukraine and 560 mGy in Belarus (Brenner et al., 2011; Zablotska et al., 2011). About 600 workers were involved in the emergency response immediately after the accident; 134 developed acute radiation syndrome resulting in 28 deaths (Kesminiene et al., 2012; Zablotska et al., 2013). Thyroid cancer was high in children aged 0—5 years at exposure, but no such increase was detected in adults. It is possible that absorbed radiation doses are much higher in children than in adults due to the small size of thyroid glands in children and the features of their metabolism. Interestingly, thyroid cancer incidence in children born after the accident was similar to background levels, which suggests that the increase in thyroid cancer near Chernobyl was mainly due to internal exposure to radioactive iodine, which has a short half-life of 8 days (Shibata et al., 2001). On the other hand, doses to tissues other than the thyroid from external or internal exposures were low (UNSCEAR, 2011).

5. Effects of low doses of ionizing radiation

Effects of ionizing radiation usually are dose dependent for acute exposures. However, the same principle or model cannot be applied to predict the risks linked to persistent exposures to small doses of ionizing radiation. Interestingly, low-dose exposures are more common than accidental exposures to higher doses. Many people are exposed to low doses of radiation from medical fields or to higher levels of natural background radiations.

5.1 Radiation workers and patients exposed to ionizing radiation

During the past decades, it has become one of the main interests to adequately evaluate exposure to ionizing radiation in different occupational groups of workers. The main focus of these studies on radiation workers was to determine the increase in cancer incidence and mortality due to occupational exposures. This is mainly to increase the radiation protection during work. Most of them are radiologists or radiotherapists, who were subjected to protracted low-dose and low-dose rate of ionizing radiation. Although many studies were performed in this regard, so far there is no statistically conclusive evidence to show detrimental effects of low-dose exposures in occupational radiation workers. This is mainly due to the fact that large cohort of exposed individuals and prolonged follow-ups are required taking into account of the long latency period in disease manifestation following low-dose radiation exposure (BEIR, 2006). Increased risks for skin cancer, leukemia, and all-cause mortality rates were reported in radiologic technologists and radiologists. Eight cohort studies of over 270,000 radiologists and radiologic technologists employed before 1950 demonstrated increased mortality rates due to leukemia (Dublin and Spiegelman, 1948). Notwithstanding, after the first recommendations on radiation protection in 1920 (dose limit equivalent to 500 mSv/year), the excess of mortality disappeared (Doll et al., 2005; Yoshinaga et al., 2004). Recent American Association of Physicists in Medicine policy states that epidemiological evidence supporting the increased cancer incidence or mortality from radiation doses below 100 mSv is inconclusive (AAPM, 2018).

On the other hand, patients are subjected to radiotherapy at high doses (between 40 and 60 Gy) during cancer treatment. In principle, surrounding tissues should not receive doses more than 0.1 Gy during radiotherapy. In some cases, depending on the cancer location, partial body irradiation requires attention for risk assessment as well as for radiation protection. On the contrary, though increased use of diagnostic X-ray examinations and fluoroscopy-guided interventional procedures in the clinical practice poses a problem for risk estimations, it is anticipated that target organs would receive small doses from these procedures (Ron, 2003). Currently, advanced imaging technologies play an essential role in the screening of asymptomatic patients. Radiation doses associated with computed tomography (CT) typically range from 1 to 20 mSv, and so far, no excess cancers were detected for doses below 100 mSv. Recently, Walsh et al. (2014) reported a direct association between radiations from CT scans and cancer induction. Moreover, a higher number of breast cancer cases were detected in women after repeated chest fluoroscopic procedures for chronic tuberculosis or scoliosis (Doody et al., 2000).

5.2 Environmental radiation exposures

Most of the background radiation is originated from a natural γ-radiation emitted by rocks, soil, and terrestrial radon. However, there is a certain percentage of background radiation coming from cosmic radiation (Hall and Giaccia, 2019). Besides, environmental radiation levels have been substantially changed by accidents in nuclear power plants as well as atomic bomb explosions. The levels of natural background radiation are different in several geographical areas of the world. Additionally, radon gas appears to be the main source of natural background radiation exposure globally as well (WHO, 2020). Generally, the average values of the effective dose rate are about 2–4 mSv/year, but regions with effective dose rate above 10 mSv/year are generally referred to as high natural background radiation areas. In areas such as Guarapari (Brazil), Kerala (India), Ramsar (Iran), and Yangjiang (China), natural background radiation can reach several hundred mSv/year (for example, in the Ramsar province, Iran, the total annual effective dose reaches 260 mSv/year). Interestingly, it has been demonstrated that the levels of natural background radiation are inversely correlated with cancer mortality (Vaiserman et al., 2018; Hart and Hyun, 2012).

6. Radiation-induced chromosome alterations

Earlier extensive cytogenetic studies on people subjected to radiation exposure in Chernobyl (Maznik and Vinnikov, 2005) and Goiânia (Natarajan et al., 1991, 1998) detected structural chromosomal aberrations immediately after exposure, and it was observed that these aberrations persisted several years postexposure. Recently, Chen et al. (2014) reported elevated levels of chromosomal aberrations in people living in the cities of Tokyo and Niigata compared with a control group, which lived far from the nuclear site at Fukushima relative. In an earlier study, we examined chromosomes in blood lymphocytes obtained from workers of the Mayak nuclear facility, Russia. Mayak workers were exposed to either γ-rays, plutonium, or both during their work period more than 60 years ago. Simple chromosome translocations were markedly higher in individuals exposed to radiation compared with the nonexposed controls, whereas stable complex chromosomal aberrations and stable intrachromosomal aberrations were observed almost exclusively in workers exposed to radioactive plutonium, but not in individuals exposed to just γ-rays (Hande et al., 2003, 2005; Mitchell et al., 2004). In Fig. 4.1A–F, different types of chromosomal aberrations induced by ionizing radiation in human blood lymphocytes are shown (Venkatesan et al., 2015).

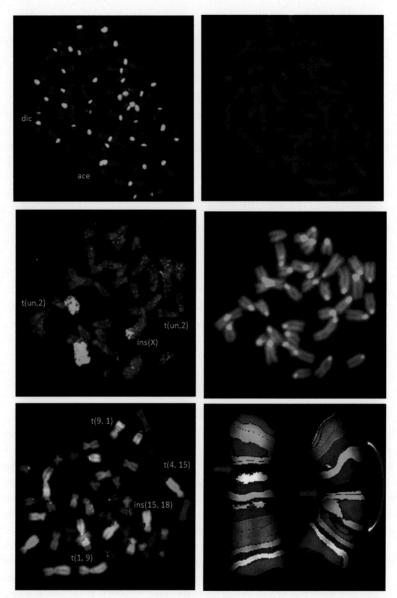

FIGURE 4.1

Direct DNA damage induced by ionizing radiation results in genetic alterations such as chromosome aberrations. Chromosome aberrations have severe implications on human health. Such chromosomal changes can be visualized by molecular cytogenetic techniques such as fluorescence in situ hybridization (FISH). (A and B) Peptide nucleic acid fluorescence in situ hybridization (PNA-FISH) was used to detect chromosome aberrations in human lymphocytes following in vitro irradiation. Cy3-telomere and FITC-

Radiation-induced direct damage can affect nuclear DNA, which is the primary target of ionizing radiation-induced damage. Radiation-induced DNA damage induces chromosome- or chromatid-type aberrations dependent on the nature and extent of DNA DSBs and stage of cell cycle at the time of exposure (Natarajan et al., 1996). Since stable aberrations, such as translocations, can be transmitted to the daughter cells, they are quite stable in the body. Such translocations remain in the system even after several years, unlike dicentrics that are eliminated from the system with time due to clearance of cells harboring dicentrics (Awa, 1990; Lucas et al., 1992). The International Atomic Energy Agency (IAEA) has recommended the use of dicentric chromosome assay in peripheral blood lymphocytes for biodosimetry purposes in accidental radiation exposures (IAEA, 1986). Natarajan et al. (1996) utilized data on chromosomal aberrations in triage scenarios for dose estimation in the radiation accident in Goiânia, Brazil. However, physiologically, lymphocyte repopulation is believed to dilute the frequency of chromosomes with dicentrics produced immediately after radiation exposure. Cells with dicentrics may be excluded from the lymphatic system in 6 months to a few years, which may depend on the radiation quality and dose received (Ramalho et al., 1995). Stable interchromosomal and intrachromosomal exchanges such as translocations and inversions were shown to persist several years after exposure unlike unstable aberrations such as dicentrics (Natarajan et al., 1998).

◄───

centromere PNA probes, depicted by the *red* and *green* signals, respectively, were used and counterstained with DAPI (*blue*). (A) Metaphase spread showing dicentric chromosomes (dic) and acentric fragments (ace). (B) The same metaphase as in (A) observed under UV filter. (C and D) Persistence of chromosome translocations in mouse splenocytes following in vivo exposure to X-rays. FISH using chromosomes 2 (*red*) and X (*green*) on metaphases of mouse splenocytes following 2-Gy irradiation with X-rays. Persistent complex chromosome aberrations involving chromosomes 2 and X, and an unpainted chromosome could be detected in mouse splenocytes 7 months postirradiation. (C) *Arrows* point to the presence of nonreciprocal translocations and insertion. (D) The same metaphase as in (C) observed under UV filter. (E and F) Multicolor fluorescence in situ hybridization (mFISH) to visualize chromosome alterations in human lymphocytes. (E) Interchromosomal aberrations (simple translocation) detected using mFISH. Reciprocal translocation involving chromosomes 1 and 9 [t(1,9) and t(9,1)] was detected in human lymphocytes after exposure to 4 Gy γ-rays. Nonreciprocal translocation between chromosomes 4 and 15 [t(4,15)] could also be seen. There is an insertion of a segment of chromosome 15 on chromosome 18 [ins(15,18)]. (F) Intrachromosomal aberrations in chromosome 5 detected using mBAND FISH. Left chromosome is normal; right chromosome shows aberrations. *Red arrows* point to the location of centromeres, and inverted *arrow* shows interarm aberrations (pericentric inversion) pointing the region of the chromosome that was inverted.

(A–F) Reproduced from Venkatesan, S., Natarajan, A.T., Hande, M.P., 2015. Chromosomal instability– mechanisms and consequences. Mutation Research: Genetic Toxicology and Environmental Mutagenesis 793, 176–184.

7. Transgenerational effects of radiation

Transgenerational effects are measured in the offspring of parents who were exposed to radiation. The mechanism behind the transmission of effect of radiation from parents to children is not completely understood. One of the earliest mechanisms proposed was that radiation-induced complex DNA DSBs may produce persistent unresolved breaks during meiosis (Limoli et al., 1997). Dubrova et al. (Dubrova et al., 2008; Baulch et al., 2014) reported that sporadic epigenetic modification in the regions of repair leads to an unstable phenotype in the progeny. A recent study showed the upregulation of a few miRNAs and maintenance of DNA methyltransferases, 2 days after radiation exposure indicating the initiation of genomic instability owing to epigenetic alterations ensuing transgenerational instability. Further investigations or studies are definitely needed to elucidate whether transgenerational effects are truly a radiation specific phenomenon (Huumonen et al., 2014). Accumulation of genetic damage as observed by abnormal traits getting propagated through generations in butterflies following radiation exposure was observed in Fukushima nuclear disaster site, and also incidence of tumor formation was observed in them as well (Taira et al., 2014). Many factors such as innate DNA repair dynamics, apoptotic death, organismal life span, and regenerative capacity, among others, play a role in the differential outcome of radiation exposure on various species (Venkatesan et al., 2015). The exact mechanisms of radiation-mediated transgenerational instability need to be studied in detail. This needs urgent attention as radiation exposures have become more frequent in the recent times.

Gardner et al. (1990) reported that the occurrence of leukemia and lymphoma in children near Sellafield nuclear plant in West Cumbria (UK) could be due to parental exposure to radiation (Nomura, 1990). In addition, high levels of dicentric chromosomes were detected in children from the contaminated areas around Chernobyl with a dose estimate of up to 0.4 Gy (Mikhalevich et al., 2000) when compared with the samples from unaffected regions. These results suggest a possibility of transmission of residual genetic/epigenetic effects from the affected parents to their children (Dubrova et al., 2008; Nomura, 1990; Barber et al., 2002; Vorobtsova, 2000). By comparing microsatellite profiles of parents exposed to ^{137}Cs radiation in Goiânia, Brazil, those of their progeny 19 years after the radiation exposure, da Cruz et al. (2008) provided a convincing evidence of the phenomenon in humans. Stable chromosome aberrations such as translocations persist in the genome; it is possible that such aberrations in germ cells would get transferred to the progeny. Indeed, epigenetic control over the locus may regulate the outcome of such expressions.

8. Epigenetics changes induced by ionizing radiations

Early studies have shown that ionizing radiation induces a DNA hypomethylation (Kalinich et al., 1989). Recently, it is considered that ionizing radiation induces not only genetic damage but also epigenetic changes (Burgio et al., 2018; Miousse et al., 2017). In this respect, it has been demonstrated in different tissues from mice that chronic exposure to low doses of radiation is a potent inducer of global DNA methylation changes in comparison with acute exposures (Kovalchuk et al., 2004). Moreover, fractionated exposure to low doses of ionizing radiation produced changes in global DNA methylation as well as on methyl transferase machinery depending on the brain tissue affected (Koturbash et al., 2016). An interesting issue is the fact that the epigenetics modifications induced by ionizing radiations are more persistent than genetic mutations. Most of the radiation-induced lesions were repaired within a month while DNA hypomethylation still persisted (Koturbash et al., 2005). Later on, several research works indicate that after the exposure to low-radiation doses, DNA hypomethylation at specific cancer-related genes can be induced (Koturbash et al., 2017).

9. Radiation protection

Main purpose of radiation protection is to reduce the stochastic effects produced by radiation, in particular cancer, and to prevent deterministic effects that are "tissue reactions." Radiological protection can be considered as "a discipline in which concepts, methods, and procedures are developed to be used for the protection of humans and the environment from the harmful effects of ionizing radiation" (Menzel and Harrison, 2012). Recommendations and practical guidance for radiological protection were developed by the International Commission on Radiological Protection or ICRP (Menzel and Harrison, 2012). ICRP (2007) projected that "the incidence of cancer or hereditary disorders may rise in direct proportion to an increase in the equivalent dose in the relevant organs and tissues below about 100 mSv"—this advice is for the purposes of radiological protection (ICRP, 2007; Wrixon, 2008).

10. Final remarks

Most of the knowledge related to the health impacts of radiation exposure in humans was generated from epidemiological studies on atomic bomb survivors. Besides, data from occupational, medical, and environmental exposures constitute the basis of the actual radiation protection system of the ICRP, which is being used by government establishments worldwide. However, absence of strong statistical power of epidemiological studies makes it difficult to estimate the risk of cancer for doses of less than 100 mGy or for moderate doses delivered at low-dose rates (BEIR, 2006). Alternatively, emerging area of high-throughput radiation biology may provide strong analytical tools to elucidate mechanisms and to assess radiation-induced cancer risk and/or other health effects with more accuracy (Preston, 2015). Earlier studies of radiation effects have mainly focused on natural science and presumably have not taken into consideration of the psychosocial aspects of radiation exposure. However, adverse psychological effects were reported among people who experienced the trauma of the Chernobyl nuclear power plant accident (UNSCEAR, 2011); in fact, this is being discussed in the case of the Fukushima Daiichi nuclear power plant accident, where psychosocial problems in Fukushima have a devastating effect on its population (Ohtsuru et al., 2015).

It is anticipated that studies on the effects of cosmic radiation exposures will be of great importance as well as space exploration, and air travel may increase in the future. These issues were discussed in a recently concluded meeting on "Living in Space: Integral Understanding of Life-regulation Mechanism from SPACE" organized by the Japanese Space Agency (JAXA, 2019) including the possible combined effects of cosmic radiation and microgravity on human health. In addition, collateral tissue damage produced by radiotherapy may produce long-term effects on cancer survivors, a topic that is beyond the scope of current review. Another important aspect to be considered in evaluating the health effects of ionizing radiation whether due to background radiation, medical radiation, or human-made sources is the genetic makeup and lifestyle of the individuals being evaluated. Genetics and lifestyle may modulate the effects of ionizing radiation in humans.

Acknowledgments

While we sincerely acknowledge the contribution of several other articles and reviews for more than a century now that have vastly improved our view of on health effects of radiation exposures in general and radiation biology in particular, we apologize for not having been able to cite all the works in this review due to space constraints. A cautionary note is that we have written this as an overview without dwelling much into individual aspects of biological effects of radiation.

References

Awa, A.A., 1990. Chromosome aberrations in A-bomb survivors, Hiroshima and Nagasaki. In: Obe, G., Natarajan, A.T. (Eds.), Chromosomal Aberrations: Basic and Applied Aspects. Springer-Verlag, Heidelberg, pp. 180−190.

AAPM, 2018. AAPM Position Statement on Radiation Risks from Medical Imaging Procedures. American Association of Physicists in Medicine Policy Statement, PP 25-C. https://www.aapm.org/org/policies/details.asp?type=PP&id=439.

Barber, R., Plumb, M.A., Boulton, E., Roux, I., Dubrova, Y.E., 2002. Elevated mutation rates in the germ line of first- and second-generation offspring of irradiated male mice. Proceedings of the National Academy of Sciences of the United States of America 99, 6877−6882.

Barcellos-Hoff, M.H., 2008. Cancer as an emergent phenomenon in systems radiation biology. Radiation and Environmental Biophysics 47, 33−38.

Baulch, J.E., Aypar, U., Waters, K.M., Yang, A.J., Morgan, W.F., 2014. Genetic and epigenetic changes in chromosomally stable and unstable progeny of irradiated cells. PLoS One 9, e107722.

BEIR, 2006. Health Risks from Exposure to Low Levels of Ionizing Radiation: BEIR VII Phase 2. The National Academies Press, Washington, DC.

Bray, F., Ferlay, J., Soerjomataram, I., Siegel, R.L., Torre, L.A., Jemal, A., 2018. Global cancer statistics 2018: GLOBOCAN estimates of incidence and mortality worldwide for 36 cancers in 185 countries. CA A Cancer Journal for Clinicians 68, 394–424.

Brenner, A.V., Tronko, M.D., Hatch, M., Bogdanova, T.I., Oliynik, V.A., Lubin, J.H., Zablotska, L.B., Tereschenko, V.P., McConnell, R.J., Zamotaeva, G.A., O'Kane, P., Bouville, A.C., Chaykovskaya, L.V., Greenebaum, E., Paster, I.P., Shpak, V.M., Ron, E., 2011. I-131 dose response for incident thyroid cancers in Ukraine related to the Chornobyl accident. Environmental Health Perspectives 119, 933–939.

Bromet, E.J., Havenaar, J.M., Guey, L.T., 2011. A 25 year retrospective review of the psychological consequences of the Chernobyl accident. Clinical Oncology 23, 297–305.

Burgio, E., Piscitelli, P., Migliore, L., 2018. Ionizing radiation and human health: reviewing models of exposure and mechanisms of cellular damage. An epigenetic perspective. International Journal of Environmental Research and Public Health 15.

Chen, Y., Zhou, P.K., Zhang, X.Q., Wang, Z.D., Wang, Y., Darroudi, F., 2014. Cytogenetic studies for a group of people living in Japan 1 year after the Fukushima nuclear accident. Radiation Protection Dosimetry 159, 20–25.

da Cruz, A.D., de Melo e Silva, D., da Silva, C.C., Nelson, R.J., Ribeiro, L.M., Pedrosa, E.R., Jayme, J.C., Curado, M.P., 2008. Microsatellite mutations in the offspring of irradiated parents 19 years after the Cesium-137 accident. Mutation Research 652, 175–179.

Doll, R., Berrington, A., Darby, S.C., 2005. Low mortality of British radiologists. British Journal of Radiology 78, 1057–1058.

Doody, M.M., Lonstein, J.E., Stovall, M., Hacker, D.G., Luckyanov, N., Land, C.E., 2000. Breast cancer mortality after diagnostic radiography: findings from the U.S. Scoliosis Cohort Study. Spine (Phila Pa 1976) 25, 2052–2063.

Drozdovitch, V., Minenko, V., Khrouch, V., Leshcheva, S., Gavrilin, Y., Khrutchinsky, A., Kukhta, T., Kutsen, S., Luckyanov, N., Shinkarev, S., Tretyakevich, S., Trofimik, S., Voilleque, P., Bouville, A., 2013. Thyroid dose estimates for a cohort of Belarusian children exposed to radiation from the Chernobyl accident. Radiation Research 179, 597–609.

Dublin, L.I., Spiegelman, M., 1948. Mortality of medical specialists, 1938–1942. Journal of the American Medical Association 137, 1519–1524.

Dubrova, Y.E., Hickenbotham, P., Glen, C.D., Monger, K., Wong, H.P., Barber, R.C., 2008. Paternal exposure to ethylnitrosourea results in transgenerational genomic instability in mice. Environmental and Molecular Mutagenesis 49, 308–311.

Folley, J.H., Borges, W., Yamawaki, T., 1952. Incidence of leukemia in survivors of the atomic bomb in Hiroshima and Nagasaki, Japan. The American Journal of Medicine 13, 311–321.

Furusawa, Y., 2014. Heavy-ion radiobiology. In: Tsujii, H., Kamada, T., Shirai, T., Noda, K., Tsuji, H., Karasawa, K. (Eds.), Carbon-ion Radiotherapy: Principles, Practices, and Treatment Planning. Springer, Japan, pp. 25–37.

Gardner, M.J., Snee, M.P., Hall, A.J., Powell, C.A., Downes, S., Terrell, J.D., 1990. Results of case-control study of leukaemia and lymphoma among young people near Sellafield nuclear plant in West Cumbria. British Medical Journal 300, 423–429.

Glasser, O., 1995. W.C. Roentgen and the discovery of the Roentgen rays. American Journal of Roentgenology 165, 1033–1040.

Hall, E.J., Giaccia, A.J., 2019. Radiobiology for the Radiologist, eighth ed. Wolters Kluwer, Philadelphia.

Hamada, N., Fujimichi, Y., 2014. Classification of radiation effects for dose limitation purposes: history, current situation and future prospects. Journal of Radiation Research 55, 629–640.

Hande, M.P., Azizova, T.V., Geard, C.R., Burak, L.E., Mitchell, C.R., Khokhryakov, V.F., Vasilenko, E.K., Brenner, D.J., 2003. Past exposure to densely ionizing radiation leaves a unique permanent signature in the genome. The American Journal of Human Genetics 72, 1162–1170.

Hande, M.P., Azizova, T.V., Burak, L.E., Khokhryakov, V.F., Geard, C.R., Brenner, D.J., 2005. Complex chromosome aberrations persist in individuals many years after occupational exposure to densely ionizing radiation: an mFISH study. Genes Chromosomes & Cancer 44, 1–9.

Hart, J., Hyun, S., 2012. Cancer mortality, state mean elevations, and other selected predictors. Dose Response 10, 58–65.

Huumonen, K., Korkalainen, M., Viluksela, M., Lahtinen, T., Naarala, J., Juutilainen, J., 2014. Role of microRNAs and DNA methyltransferases in transmitting induced genomic instability between cell generations. Frontiers in Public Health 2, 139.

IAEA, 1986. Biological Dosimetry: Chromosomal Aberrations Analysis for Dose Assesment. International Atomic Energy Agency, Vienna.

ICRP, 2007. The 2007 recommendations of the international commission on radiological protection. ICRP publication 103. Annals of the ICRP 37, 1–332.

JAXA, 2019. In: Living in Space: Integral Understanding of Life-Regulation Mechanism From "SPACE", International Symposium on Living in Space. https://living-in-space.sakura.ne.jp/en/symposium/mailform/.

Kalinich, J.F., Catravas, G.N., Snyder, S.L., 1989. The effect of gamma radiation on DNA methylation. Radiation Research 117, 185–197.

Kamiya, K., Ozasa, K., Akiba, S., Niwa, O., Kodama, K., Takamura, N., Zaharieva, E.K., Kimura, Y., Wakeford, R., 2015. Long-term effects of radiation exposure on health. Lancet 386, 469–478.

Kesminiene, A., Evrard, A.S., Ivanov, V.K., Malakhova, I.V., Kurtinaitise, J., Stengrevics, A., Tekkel, M., Chekin, S., Drozdovitch, V., Gavrilin, Y., Golovanov, I., Kryuchkov, V.P., Maceika, E., Mirkhaidarov, A.K., Polyakov, S., Tenet, V., Tukov, A.R., Byrnes, G., Cardis, E., 2012. Risk of thyroid cancer among Chernobyl liquidators. Radiation Research 178, 425–436.

Koturbash, I., Pogribny, I., Kovalchuk, O., 2005. Stable loss of global DNA methylation in the radiation-target tissue—a possible mechanism contributing to radiation carcinogenesis? Biochemical and Biophysical Research Communications 337, 526–533.

Koturbash, I., Kutanzi, K., Hendrickson, K., Rodriguez-Juarez, R., Kogosov, D., Kovalchuk, O., 2008. Radiation-induced bystander effects in vivo are sex specific. Mutation Research 642, 28–36.

Koturbash, I., Jadavji, N.M., Kutanzi, K., Rodriguez-Juarez, R., Kogosov, D., Metz, G.A.S., Kovalchuk, O., 2017. Fractionated low-dose exposure to ionizing radiation leads to DNA damage, epigenetic dysregulation, and behavioral impairment. Environmental Epigenetics 2, 1–13.

Kovalchuk, O., Burke, P., Besplug, J., Slovack, M., Filkowski, J., Pogribny, I., 2004. Methylation changes in muscle and liver tissues of male and female mice exposed to acute and chronic low-dose X-ray-irradiation. Mutation Research 548, 75–84.

Land, C.E., Tokunaga, M., Tokuoka, S., Nakamura, N., 1993. Early-onset breast cancer in A-bomb survivors. Lancet 342, 237.

Limoli, C.L., Kaplan, M.I., Phillips, J.W., Adair, G.M., Morgan, W.F., 1997. Differential induction of chromosomal instability by DNA strand-breaking agents. Cancer Research 57, 4048−4056.

Little, J.B., 2003a. Genomic instability and bystander effects: a historical perspective. Oncogene 22, 6978−6987.

Little, M.P., 2003b. Risks associated with ionizing radiation. British Medical Bulletin 68, 259−275.

Lucas, J.N., Poggensee, M., Straume, T., 1992. The persistence of chromosome translocations in a radiation worker accidently exposed to tritium. Cytogenetics and Cell Genetics 60, 255−256.

Maznik, N.A., Vinnikov, V.A., 2005. The retrospective cytogenetic dosimetry using the results of conventional chromosomal analysis in Chernobyl clean-up workers. Radiatsionnaia Biologiia, Radioecologiia/Rossiiskaia Akademiia Nauk 45, 700−708.

Menzel, H.G., Harrison, J., 2012. Effective dose: a radiation protection quantity. Annals of the ICRP 41, 117−123.

Mikhalevich, L.S., De Zwart, F.A., Perepetskaya, G.A., Chebotareva, N.V., Mikhalevich, E.A., Tates, A.D., 2000. Radiation effects in lymphocytes of children living in a Chernobyl contaminated region of Belarus. International Journal of Radiation Biology 76, 1377−1385.

Miousse, I.R., Kutanzi, K.R., Koturbash, I., 2017. Effects of ionizing radiation on DNA methylation: from experimental biology to clinical applications. International Journal of Radiation Biology 93, 457−469.

Mitchell, C.R., Azizova, T.V., Hande, M.P., Burak, L.E., Tsakok, J.M., Khokhryakov, V.F., Geard, C.R., Brenner, D.J., 2004. Stable intrachromosomal biomarkers of past exposure to densely ionizing radiation in several chromosomes of exposed individuals. Radiation Research 162, 257−263.

Mutscheller, A., 1950. The biological significance of the roentgen. Journal of Investigative Dermatology 14, 471−481.

Natarajan, A.T., Vyas, R.C., Wiegant, J., Curado, M.P., 1991. A cytogenetic follow-up study of the victims of a radiation accident in Goiania (Brazil). Mutation Research 247, 103−111.

Natarajan, A.T., Balajee, A.S., Boei, J.J., Darroudi, F., Dominguez, I., Hande, M.P., Meijers, M., Slijepcevic, P., Vermeulen, S., Xiao, Y., 1996. Mechanisms of induction of chromosomal aberrations and their detection by fluorescence in situ hybridization. Mutation Research 372, 247−258.

Natarajan, A.T., Santos, S.J., Darroudi, F., Hadjidikova, V., Vermeulen, S., Chatterjee, S., Berg, M., Grigorova, M., Sakamoto-Hojo, E.T., Granath, F., Ramalho, A.T., Curado, M.P., 1998. [137]Cesium-induced chromosome aberrations analyzed by fluorescence in situ hybridization: eight years follow up of the Goiania radiation accident victims. Mutation Research 400, 299−312.

Nomura, T., 1990. Of mice and men? Nature 345, 671.

Nussbaum, R.H., Kohnlein, W., 1994. Inconsistencies and open questions regarding low-dose health effects of ionizing radiation. Environmental Health Perspectives 102, 656−667.

Nussbaum, R.H., Kohnlein, W., 1995. Health consequences of exposures to ionizing radiation from external and internal sources: challenges to radiation protection standards and biomedical research. Medicine and Global Survival 2, 199−213.

Ohtsuru, A., Tanigawa, K., Kumagai, A., Niwa, O., Takamura, N., Midorikawa, S., Nollet, K., Yamashita, S., Ohto, H., Chhem, R.K., Clarke, M., 2015. Nuclear disasters and health: lessons learned, challenges, and proposals. Lancet 386, 489−497.

Ozasa, K., Grant, E.J., Kodama, K., 2018. Japanese legacy cohorts: the life span study atomic bomb survivor cohort and survivors' offspring. Journal of Epidemiology 28, 162−169.

Pierce, D.A., Sharp, G.B., Mabuchi, K., 2005. Joint effects of radiation and smoking on lung cancer risk among atomic bomb survivors. Radiation Research 163, 694−695.

Pouget, J.P., Mather, S.J., 2001. General aspects of the cellular response to low- and high-LET radiation. European Journal of Nuclear Medicine 28, 541−561.

Preston, R.J., 2015. Integrating basic radiobiological science and epidemiological studies: why and how. Health Physics 108, 125−130.

Radvanyi, P., Bordry, M., 1984. La radioactivite artificielle et son histoire, Seuil. CNRS, Paris.

Radvanyi, P., Bordry, M., 1988. Histoires D'atomes, Belin, Paris.

Ramalho, A.T., Curado, M.P., Natarajan, A.T., 1995. Lifespan of human lymphocytes estimated during a six year cytogenetic follow-up of individuals accidently exposed in the 1987 radiological accident in Brazil. Mutation Research 331, 47−54.

Ron, E., 2003. Cancer risks from medical radiation. Health Physics 85, 47−59.

Shibata, Y., Yamashita, S., Masyakin, V.B., Panasyuk, G.D., Nagataki, S., 2001. 15 years after Chernobyl: new evidence of thyroid cancer. Lancet 358, 1965−1966.

Taira, W., Nohara, C., Hiyama, A., Otaki, J.M., 2014. Fukushima's biological impacts: the case of the pale grass blue butterfly. Journal of Heredity 105, 710−722.

UNSCEAR, 2011. Report of the United Nations Scientific Committee on the Effects of Atomic Radiation 2010, Vienna, Austria.

Upton, A.C., Shore, R.E., Harley, N.H., 1992. The health effects of low-level ionizing radiation. Annual Review of Public Health 13, 127−150.

Vaiserman, A., Koliada, A., Zabuga, O., Socol, Y., 2018. Health impacts of low-dose ionizing radiation: current scientific debates and regulatory issues. Dose Response 16, 1559325818796331.

Venkatesan, S., Natarajan, A.T., Hande, M.P., 2015. Chromosomal instability−mechanisms and consequences. Mutation Research: Genetic Toxicology and Environmental Mutagenesis 793, 176−184.

Vorobtsova, I.E., 2000. Irradiation of male rats increases the chromosomal sensitivity of progeny to genotoxic agents. Mutagenesis 15, 33−38.

WHO, 2020. World Health Organisation: Ionizing Radiation. Environmental Radiation. Accessed 2020. http://www.who.int/ionizing_radiation/env/en/.

Walsh, L., Shore, R., Auvinen, A., Jung, T., Wakeford, R., 2014. Risks from CT scans−what do recent studies tell us? Journal of Radiological Protection 34, E1−E5.

Wrixon, A.D., 2008. New ICRP recommendations. Journal of Radiological Protection 28, 161−168.

Yasumura, S., Hosoya, M., Yamashita, S., Kamiya, K., Abe, M., Akashi, M., Kodama, K., Ozasa, K., 2012. Study protocol for the Fukushima health management Survey. Journal of Epidemiology 22, 375−383.

Yoshinaga, S., Mabuchi, K., Sigurdson, A.J., Doody, M.M., Ron, E., 2004. Cancer risks among radiologists and radiologic technologists: review of epidemiologic studies. Radiology 233, 313−321.

Zablotska, L.B., Ron, E., Rozhko, A.V., Hatch, M., Polyanskaya, O.N., Brenner, A.V., Lubin, J., Romanov, G.N., McConnell, R.J., O'Kane, P., Evseenko, V.V., Drozdovitch, V.V., Luckyanov, N., Minenko, V.F., Bouville, A., Masyakin, V.B., 2011.

Thyroid cancer risk in Belarus among children and adolescents exposed to radioiodine after the Chornobyl accident. British Journal of Cancer 104, 181–187.

Zablotska, L.B., Bazyka, D., Lubin, J.H., Gudzenko, N., Little, M.P., Hatch, M., Finch, S., Dyagil, I., Reiss, R.F., Chumak, V.V., Bouville, A., Drozdovitch, V., Kryuchkov, V.P., Golovanov, I., Bakhanova, E., Babkina, N., Lubarets, T., Bebeshko, V., Romanenko, A., Mabuchi, K., 2013. Radiation and the risk of chronic lymphocytic and other leukemias among chornobyl cleanup workers. Environmental Health Perspectives 121, 59–65.

Nuclear plant severe accidents: challenges and prevention

Wison Luangdilok, PhD, MS, BS [1,2], **Peng Xu** [3]

[1]*President, H2Technology LLC, Westmont, IL, United States;* [2]*Fauske & Associates LLC, Burr Ridge, IL, United States;* [3]*Idaho National Laboratory, Idaho Falls, ID, United States*

1. Introduction

A severe nuclear reactor accident is the only source of risk to the public from an operating light water reactor power plant (Sehgal, 2012). This public risk of nuclear power can be reduced by prevention, mitigation, and management of severe accidents. The idea of prevention and management of severe accidents was the product of the lessons learned from the TMI-2 (Three Mile Island Unit 2) accident that occurred on March 28, 1979. Apparently, the lessons learned from the TMI-2 accident were not enough to prevent a severe accident at the Fukushima Daiichi nuclear power station on March 11, 2011, which was triggered by the largest scale earthquake and tsunami that were ever recorded in Japan. Prior to the Fukushima accident, severe accidents (with a very low probability of occurrence and a very high dose consequence) had not been included in a set of accident scenarios called "design-basis accidents (DBAs)" that are required for licensing in most or all countries. This type of safety standard in the licensing of light water reactor nuclear power plants left open a very small chance of a severe accident that, if it ever happens, there would be a big consequence that comes with it. No one had ever thought that the Fukushima accident would be the unfortunate one.

Several investigations by various organizations on the cause of the Fukushima accident were initiated in either 2011 or 2012 and completed. The lessons learned for accident prevention and safety improvement were studied and implementations of the lessons learned have resulted in the new post-Fukushima safety requirements, for example, in Japan (NRA, 2013a, 2013b) and the United States (NRC, 2012a, 2012c, 2013), and have also led to an update of the International Atomic Energy Agency (IAEA) (IAEA, 2016, 2019) safety standards to assure safety of the commercial light water reactors worldwide against severe accidents. At the time of this writing, implementation of the new safety requirements to existing nuclear plants have been ongoing in Japan, while that of the US plants has been completed.

This chapter discusses (1) the updated understanding of the Fukushima accident from the perspective of looking back 9 years later, with additional insight from accident analyses that were not available at earlier years; (2) the lessons learned (including insights relevant to severe accident prevention, mitigation, and management) as summarized by the study of the US National Academy of Sciences (NAS) (NAS, 2014); (3) the development of the new post-Fukushima safety regulations for coping with beyond DBAs; and (4) the implementation of the safety regulations in the nuclear power plant restart program in Japan and implementation to plant safety upgrade in the United States. Also included in the discussion is the reform in the regulatory system in Japan, which was identified as the root cause of the Fukushima accident.

2. Fukushima Daiichi accidents and insights from the accident analyses

On March 11, 2011, at 2:46 p.m., a 9.1-magnitude earthquake struck the area off the northeastern coast of Japan's main island. All reactors (except for Unit 4) at the Fukushima Daiichi site automatically and successfully scrammed as designed within about 1.5 minutes. Units 1, 2, and 3 were in operation at full power at that time, while Units 4, 5, and 6 were in periodic inspection outage, with fuel being unloaded only in Unit 4. Both Units 5 and 6 had fuel loaded. Unit 5 was performing a reactor vessel pressure leakage test, while Unit 6 was in a cold shutdown condition. Then 50 minutes after the earthquake, at 3:36 p.m., the worst nuclear accidents that ever happened to light water reactors at the Fukushima Daiichi nuclear plant site began with the arrival of the 14-m-high tsunami waves that hit the Fukushima Daiichi nuclear power station site and caused an extended station blackout (a simultaneous loss of all AC and DC power) over several days (off-site AC power was not restored until March 20 for Units 1 and 2 and until March 22 for Unit 3.). Emergency diesel generators (EDGs) and switchboards in the basement floor of the reactor buildings were flooded and disabled, except for one EDG at Unit 6. During this time, without power to run any emergency pumps for injecting water into the reactors on time, three reactors at Units 1, 2, and 3 eventually underwent a core meltdown. Unit 4 had no meltdown because the fuel was not in the reactor. Units 5 and 6 were safely brought to a cold shutdown. One of the EDGs at Unit 6 was fortunately installed at a higher ground, and as a result, its functions were not lost when the site was hit by the tsunami. Unit 5 received power from an EDG at Unit 6. Details of the events are provided in the report submitted to the IAEA by the Japanese Government (2011).

After the tsunami hit the Fukushima Daiichi site, Units 2 and 3 did not lose their DC power supply immediately like Unit 1. Because of this delayed loss of DC power in Units 2 and 3, the reactor core isolation cooling (RCIC) system was available and started in Unit 2 and the RCIC and the high-pressure core injection (HPCI) system were available for operation in Unit 3. The RCIC and HPCI systems were a core cooling system of the same design except that HPCI was about seven times larger than RCIC in flow capacities. The RCIC and HPCI systems ran on turbine pumps driven by high-pressure steam from inside the reactors. Their availability had delayed the reactors in Units 2 and 3 from overheating as long as the DC power was available. Both RCIC and HPCI systems required DC power to actuate valves, but AC power was not required to run pumps. On the other hand, the isolation condenser (IC) in Unit 1 required both AC and DC power for operation of flow control valves to remain open. AC power was required to keep shutoff valves inside the primary containment open. DC power was required to open shutoff valves outside the primary containment. The IC in Unit 1 failed almost immediately when the tsunami hit the site. As a result, the core cooling systems were available at varying times during the accident for individual units. The IC for Unit 1 was available for a short period before the tsunami arrived. For Unit 2, the RCIC system was activated 3 minutes after the arrival of tsunami but 2 minutes before the loss of DC power. So it was activated successfully and ran for almost 3 days. For Unit 3, only AC and partial DC power was lost initially. The RCIC and the HPCI systems were available sequentially for Unit 3 until batteries were depleted in about 1 day and a half.

2.1 Event progression at unit 1

As the investigation and analyses have shown, for Unit 1, the gravity-driven IC core cooling system was started and stopped four times to prevent overcooling before the arrival of the tsunami at 3:36 p.m. of March 11, 2011. Although the flow in the IC was gravity-driven, its flow control valves were all closed once the AC and DC power was lost, causing the system to be shut down. After that, it was never restarted again because of the loss of AC and DC power (i.e., the beginning of total station blackout) that began just 1 minute after the arrival of the tsunami. According to the Nuclear Energy Agency (NEA, 2015) analyses, the core was uncovered around 5:30 to 6:00 p.m. of March 11. The Tokyo Electric Power Company (TEPCO, 2017) suggests that by around 7 p.m. of March 11, the core began melting (Fig. 5.1). By around 11 p.m., some molten core relocated to the lower plenum. By around 1:30 a.m. of March 12, a larger relocation of molten core into the lower plenum occurred again. By this time, it was already too late to save the core. Any water injection into the reactor would be for cooling and arresting the molten core from causing further attack on the reactor vessel lower head. Eventually, fresh water was managed to be injected by the emergency crew from a fire engine at a much later time, at 4:00 a.m. on March 12, more than 12 hours after the IC was last manually operated.

Fire trucks and a fire brigade played a critical role in injecting water into the reactor amid high radiation. A piping route analysis by TEPCO shows that the line that was used for injecting water from the fire engine into the reactor was not a dedicated line to the reactor (Lesson learned: It is to be expected that most pre-Fukushima nuclear plants would not have such a dedicated line for emergency injection. This became an insight for improving severe accident prevention and mitigation measures.). There were branching pipes that would have diverted the water pumped from the fire engine to the condensate storage tank (CST) and the main condenser. Therefore the amount of water injected into the reactor was significantly less than the amount from the fire engine. Based on the abrupt changes in the on-site

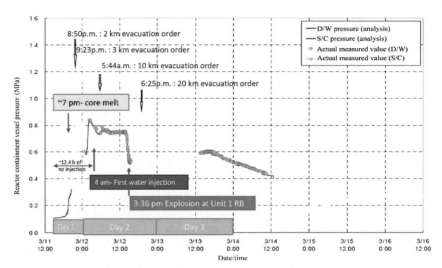

FIGURE 5.1

Major reactor events on the primary containment vessel pressure plot during an accident at Fukushima Daiichi Unit 1.

Modified from Japanese Government, Report to the IAEA Ministerial Conference on Nuclear Safety, the Accident at TEPCO's Fukushima Nuclear Power Station, 2011. https://japan.kantei.go.jp/kan/topics/201106/iaea_ houkokusho_e.html.

dose data that coincide with the change in the primary containment vessel (PCV) pressure, the reactor vessel lower head was estimated by TEPCO to fail by around 6 a.m. of March 12 (TEPCO, 2017). There were a total of four evacuation orders during days 1 and 2 of the accident. All orders were associated with the radiologic release caused by the core meltdown of Unit 1: (1) the 2-km radius evacuation order at 8:50 p.m. of March 11, (2) the 3-km radius evacuation order at 9:23 p.m. of March 11, (3) the 10-km evacuation order at 5:44 a.m. of March 12, and (4) the 20-km radius evacuation order at 6:25 p.m. of March 12 (Fig. 5.1). By this time, the reactor core of Units 2 and 3 was still intact. There were a total of three explosions as will be discussed later. The first explosion that occurred at 3:36 p.m. of March 12 was associated with Unit 1. The fourth evacuation order was announced about 3 hours after the Unit 1 explosion.

2.2 Event progression at unit 3

The next reactor that suffered a core meltdown after Unit 1 was Unit 3. After the tsunami arrived at the site at 3:36 p.m. of March 11, Unit 3 lost its on-site AC power due to the tsunami flooding, but it did not lose its on-site DC power (batteries) for about 35 hours later when the battery power was used up. This made the accident events at Unit 3 different from those in the other units. The DC power was available for the operation of the RCIC system and the HPCI system, both of which did not require AC power for operation, but the DC power was needed for controlling the RCIC and HPCI system valves. Both the RCIC and HPCI systems were driven by a steam turbine pump using high-pressure steam from the reactor vessel produced by the decay heat. The RCIC pump was manually started at 4:03 p.m. The RCIC ran for about 19.5 hours and automatically tripped due to high pressure in the suppression chamber at 11:36 a.m. of March 12. Then, about 1 hour later, at 12:35 p.m. of March 12, the HPCI system automatically started and ran for about 14 hours until 2:42 a.m. of March 13 when it was manually stopped. At this time, according to the NEA (2015) analyses, the water level in the reactor core was between 1 meter below the top of active fuel (TAF) and about the level of TAF. After this, a core damage was expected within 2−3 hours by the NEA (2015) analyses. The reactor experienced a period of no cooling for at least six and a half hours before the first injection of fresh water into the reactor was attempted from a fire engine at 9:25 a.m. of March 13 (Fig. 5.2). Prior to the freshwater injection, the reactor depressurization needed to occur. The reactor was manually depressurized at about 9:08 a.m. of March 13. The first injection continued for about 2 hours until 12:20 p.m. of March 13 due to a depletion of freshwater inventory. The source of water was then switched to seawater.

About 1 hour after the termination of the first injection, the second injection of seawater began at 1:12 p.m. of March 13 and continued for nearly 12 hours until 1:10 a.m. of March 14 when the source pit ran out of water. It then took almost 2 hours before the third injection with seawater resumed at 3:20 a.m. of March 14, which continued for 7 hours and 41 minutes until the operation was halted by

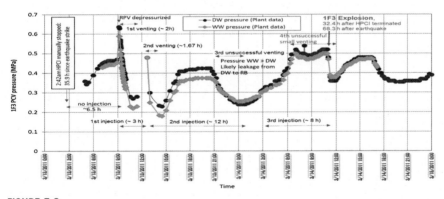

FIGURE 5.2

Major reactor events after the termination of reactor core isolation cooling and high-pressure core injection (HPCI) pumps on the primary containment vessel (PCV) pressure plot during an accident at Fukushima Daiichi Unit 3 (1F3). *DW*, drywell; *RPV*, reactor pressure vessel; *WW*, wet well.

From W. Luangdilok, The explosions at Fukushima Daiichi unit 3 and unit 4 and implications on the evaluation of 1F3 accident, Nuclear Engineering and Design 362 (2020), 110536. https://doi.org/10.1016/j.nucengdes. 2020.110536.

the explosion from the top floor of the reactor building of Unit 3 at 11:01 a.m. of March 14. After four and a half hours of the explosion, the injection of seawater was resumed at 3:30 p.m. of March 14 and continued with some interruption (for about 30 minutes to 1 hour) in between until 9:14 p.m. of March 14 when it was halted again. Seawater injection was resumed again at 2:30 a.m. of March 15.

It is noted that during the first injection of water, venting of the PCV was also manually performed for about 2 hours from 9:20 a.m. to 11:17 a.m., and later the second venting was performed between 12:30 p.m. and 2:10 p.m. The significance of the PCV venting was that the hydrogen gas that was generated from the core meltdown in Unit 3 was transferred to Unit 4 during the venting, and later on, this hydrogen gas exploded at Unit 4.

2.3 Event progression at unit 2

The Unit 2 reactor was the last reactor that suffered a core meltdown. For Unit 2, after the tsunami hit the site at 3:36 p.m. of March 11, the RCIC system, which was driven by high pressure steam from the reactor vessel, was manually started at 3:39 p.m. while the DC power was still available. The loss of both AC and DC power (i.e., the beginning of total station blackout) at Unit 2 due to flooding by the tsunami did not occur until 2 minutes later at 3:41 p.m. From here on the magical moment for Unit 2 started. The RCIC system then surprisingly ran and maintained a stable reactor water level in an uncontrolled mode for almost 3 days (Fig. 5.3). The nearly 3-day-long performance of RCIC was far beyond design specifications and even an expert's expectation (*This became a good insight showing how such equipment can save the reactor for an extended period without relying on external AC power. As discussed by* Rempe et al. (2019), *an extensive testing of RCIC*

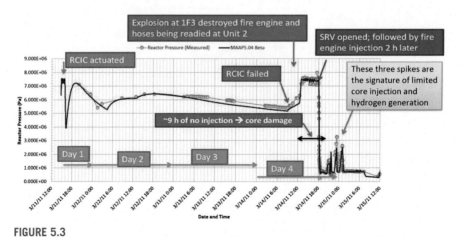

FIGURE 5.3

Major reactor events on the reactor pressure plot during an accident at Fukushima Daiichi Unit 2. *1F3*, Unit 3; *RCIC*, reactor core isolation cooling; *SRV*, safety relief valve.

performance to define ultimate operating limits of RCIC has been ongoing in the United States so that plant operators can take advantage of it.).

Eventually, the RCIC system became degraded and stopped running in the morning of March 14 between 9:00 a.m. and 12:00 p.m. based on the rise of the reactor pressure (Fig. 5.3) and the fall of water level as a sign of core heat-up due to lack of cooling. At this time, the water level in the reactor core was at least 5 m above the TAF. It would have taken another ∼5 hours before the water level dropped to the TAF and another hour before core damage began (NEA, 2015).

After the loss of RCIC, there was a period of about 9 hours before injection of water from a fire engine was successful at around 8 p.m. of March 14 (Fig. 5.3). Unfortunately, during this time the fuel had already begun to melt. The injection from the fire engine was too late. Fire trucks and a fire brigade played a critical role in injecting water into the reactor amid high radiation and concrete debris from the explosion at Unit 1 that occurred at 3:36 p.m. of March 12 and later on fresh concrete debris from the explosion of Unit 3 that occurred at 11:01 a.m. of March 14. With two successive explosions from the nearby Units 1 and 3, the difficulty faced by the fire brigade cannot be understated. The explosion at Unit 3 also destroyed the fire engine and hoses being readied for Unit 2. Although it is an academic exercise to imagine that if the Unit 3 explosion occurred at a later time, the emergency fire engine injection to prevent the meltdown could have been successful at Unit 2.

Similar to the case for Unit 1, a piping route analysis by TEPCO shows that the line that was used for injecting water from the fire engine into the reactor was not a dedicated line to the reactor. There were branching pipes that would have diverted the water pumped from the fire engine to the CST and the main condenser. Therefore the amount of water injected into the reactor was significantly less than the amount from the fire engine (TEPCO, 2017). Based on an abrupt increase of radiation dose in the PCV monitored by the containment air monitoring system, TEPCO estimated that the reactor vessel lower head failure likely occurred between 1:00 and 3:00 p.m. of March 15.

2.4 Explosions

Altogether, there were three separate explosions initiated from the top floor of the reactor building of Unit 1 (1F1), Unit 3 (1F3), and Unit 4 (1F4) at 3:36 p.m. on March 12, 2011; 11:01 a.m. on March 14, 2011; and 6:14 a.m. on March 15, 2011, respectively (Japanese Government, 2011). There was no explosion at Unit 2 (1F2). The explosions at Units 1 and 3 were caught on video cameras (with no sound recorded) by television stations from a far distance that gave the overall picture of the dynamics of the explosions. If there were no such videos, it would have been impossible to know that the explosion dynamics of 1F1 and 1F3 was very different (Fig. 5.4). The explosion dynamics of 1F4 remains unknown, as there was no video capture of the 1F4 explosion at the time it happened. One interesting twist is that, according to the TEPCO investigation, the migration of hydrogen in vented gas from Unit 3 to Unit 4 through the connected vent lines to the same shared

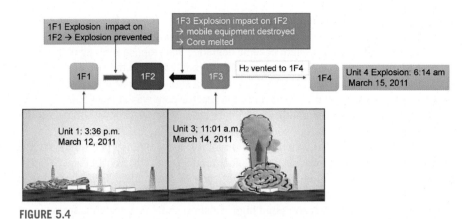

FIGURE 5.4

Three hydrogen explosions during Fukushima Daiichi accident at Units 1, 3, and 4.

From W. Luangdilok, The explosions at Fukushima Daiichi unit 3 and unit 4 and implications on the evaluation of 1F3 accident, Nuclear Engineering and Design 362 (2020) (2020), 110536. https://doi.org/10.1016/j. nucengdes.2020.110536.

vent stack during the venting of Unit 3 was responsible for supplying fuel to the 1F4 explosion. Furthermore, it is believed that the potential explosion at 1F2 was averted because of two reasons. First, the pressure wave from the 1F1 explosion caused the 1F2 reactor building top-floor blowout panel to open allowing accumulated hydrogen to vent out of the building (Rempe et al., 2019). Second, due to the unique situation at 1F2, the steam, which was generated from the 1F2 reactor building bottom-floor torus room that was partially flooded by the tsunami (NEA, 2015), helped render inert the supposedly flammable mixture in the top floor of the reactor building.

Based on the current understanding and the pattern of the occurrences, the explosion from the top floor of the reactor building has become a unique signature of a severe reactor core meltdown accident in a boiling water reactor (BWR) nuclear power plant of this type of design. During a reactor core meltdown and later on a core-concrete interaction following the reactor lower head failure at Units 1, 2, and 3, hydrogen gas was produced inside the reactor core from a high-temperature oxidation of zirconium alloy, stainless steel, and boron carbide with steam. Zirconium alloy is typically used in the BWR nuclear fuel assemblies as a cladding material for fuel pellets and as a wall material for the channel box. Stainless steel is used as a sheathing material for neutron absorber while boron carbide ceramic is used as a neutron-absorbing material in the control blade. For Unit 1, the potential source of hydrogen generation could be as high as 1730 kg, whereas the potential source for Units 2 and 3 could be even close to 3000 kg if all the above-mentioned materials were to be completely oxidized during the progression of the accident. The potential source of hydrogen generation was higher in Units 2 and 3 because 23they were a larger reactor with a rated electric output of 784 MW, whereas the rated electric output power for Unit 1 was more than 40% smaller at 460 MW. According to a recent estimate, the explosion at Unit 3 was about 11 times bigger than the explosion at Unit 1 in terms of the amount of hydrogen burned (Luangdilok, 2020).

It is important that a hydrogen explosion for such a reactor design can be prevented. All the three explosions at Fukushima Daiichi Units 1, 3, and 4 occurred at the top floor (refueling floor) of the reactor buildings. Explosions pose risk to the safety of a spent fuel pool (SFP), which is on the same top floor (refueling floor) of the reactor building. As a lesson learned from this experience, post-Fukushima BWR plants in Japan have been modified to be able to vent gases from the refueling floor of the reactor building through vent panels placed on the ceiling to avert an explosion during a severe accident. In addition, most BWR plants in Japan also have passive autocatalytic recombiners installed inside the reactor building to keep the reactor building hydrogen concentrations low during an accident (Mizokama, 2020).

2.5 Unit 1 explosion and spent fuel pool conditions

The explosion at Unit 1 heavily damaged the fifth floor (refueling floor) of the reactor building but did no damage to the floors below, except for a limited damage observed near the equipment hatch opening in the southwest corner of the fourth floor. The wall panels on all sides of the fifth floor were completely blown out, but the steel girders that supported the panels remained intact. The refueling machine and overhead crane apparently were found to be intact (NAS, 2016). However, the wreckage from the explosion prevented direct access to the SFP and direct visual observations of the condition of the SFP, which was also situated on the fifth floor of the reactor building.

Initially, several approaches were considered for adding water to the Unit 1 SFP including (1) sending personnel into the reactor building to connect fire hoses to a SFP cooling line, (2) dropping ice or water from helicopters into the pool, and (3) spraying water into the pool using a fire truck equipped with a ladder. All these approaches were deemed to be impractical or to pose unacceptable risks to plant personnel (NAS, 2016). Eventually, fresh water was pumped onto the collapsed roof structure and some of it ran off into the pool for the first time on March 31, 2011, 20 days after the accident started. The amount of water that reached the pool was uncertain because the pool could not be observed visually. This is in sharp contrast to Unit 2 that did not experience a hydrogen explosion. Water was added to the SFP in Unit 2 on March 20, 2011, just 9 days since the accident started.

The hydrogen explosion at Unit 1 not only raised a concern about the unobservable conditions (even with a video taken by a flyover helicopter during an accident) of the SFP but also significantly hindered other recovery efforts. Debris from the explosion damaged both the power lines that had been laid down at Unit 2 and the power lines being readied at Unit 3. This adversely impacted work being done to restore power at both Units 2 and 3.

2.6 Unit 3 explosion and spent fuel pool conditions

On the other hand, the explosion at Unit 3 caused much more extensive damage than the Unit 1 explosion (Rempe et al., 2019). In fact, the Unit 3 explosion damaged the fire engines and hoses being readied at Unit 2 to the extent that they could no longer be used. Not only the top fifth floor was severely damaged but also the fourth floor and nearby buildings were largely damaged. The overhead crane dropped onto the floor of the fifth floor. The roof of the turbine building experienced some damage. The top of the two-story Radwaste Building adjacent to the Unit 3 reactor building also experienced some damage. The explosion introduced a substantial amount of debris, building structural components, and equipment, including the fuel handling machine, into the SFP. These materials were resting in the bottom of the pool and on top of the racks. Similar to the Unit 1 situation, the wreckage from the explosion prevented direct access to the SFP and direct visual observations of the condition of the SFP. The hydrogen explosion at Unit 3 had raised a concern about the unobservable conditions of the SFP. Because of the wreckage the condition of the SFP could not be observed even with a video or photo taken by a flyover helicopter. There were significant difficulties faced by plant operators in attempting to add water using helicopters and fire trucks. The helicopter water drops were unsuccessful, and the fire truck sprays were ineffective. An attempt to inject seawater through an improvised connection to the fuel pool cooling system also failed. It was not until April 26, 2011, 25 days since the accident started, when water was successfully injected using a concrete pump truck. Important insights are knowing the conditions of the SFP (NAS, 2016).

2.7 Unit 4 explosion and spent fuel pool conditions

The immediate aftermath of the explosion at Unit 4 had a nervous moment of its own. Most of the roof slab and walls on the fifth floor were blown off leaving only the frame structure of the pillar and beams. Most walls on the fourth floor and some exterior surfaces on the third floor were damaged. Debris from the explosion were scattered on the refueling deck and into the SFP. Unit 4 reactor was shut down for maintenance, and large-scale repairs of RPV internals were in progress. The reactor had no fuel in the core. All the 548 fuel assemblies had been removed from the reactor and placed into contiguous racks in the SFP. The SFP at Unit 4 had fresh spent fuel with much higher decay heat than the SFPs at Units 1, 2, and 3. Not known at the time of the explosion at Unit 4 on March 15, 2011, was that the explosion was fueled by the hydrogen gas vented from Unit 3. Most off-site observers in Tokyo at the time were misled to think that the SFP at Unit 4 was in trouble and was responsible for the source of hydrogen for the explosion. From the TEPCO visual observations and a video recording made during the helicopter flyover on the next day (March 16, 2011), it was showed that the Unit 4 pool contained water above the top of the fuel racks. It was not until March 20, 2011, when plant operators attempted to add water to the SFP using water cannons and fire truck sprays with limited effectiveness. From March 22, 2011, the delivery method was switched to a concrete pump truck. For more discussions on the safety of SFPs, readers are suggested to consult the phase 2 study of the lessons learned from the Fukushima nuclear accident by the NAS (NAS, 2016).

2.8 Remarks on the accident causes

First, the Fukushima Daiichi nuclear accident was an extended total station blackout due to a loss of all AC power and DC power. The earthquake destroyed the power lines to the power station site leading to the loss of off-site AC power. Then the tsunami flooding of the reactor building basements led to the loss of on-site AC and DC power. The flooding resulted in the damage of EDGs that were located in the basement leading to the loss of on-site AC power. The loss of on-site DC power was caused by the flooding of the power distribution panel at Units 1 and 2 and by battery depletion at Unit 3. In its July 2012 report, the National Diet Independent Investigation Commission wrote that the commission believed there was a possibility that the earthquake damaged the equipment necessary for ensuring safety and that there was also a possibility that a small-scale loss-of-coolant accident (LOCA) occurred in Unit 1 (NAIIC, 2012). The Investigation Commission suggested that these points should be examined further. Regarding these assertions, a confirmation whether this was the case or not can only be done by a visual robotic examination of the piping components in question inside the highly radioactive drywell. Subsequent accident analyses (TEPCO, 2013, 2015, 2017) as well as PCV examinations by TEPCO thus far (through 2019, as summarized by Rempe et al. (2019)) have not shown any evidence that supports such possibilities yet. Another evidence that the 9.1-magnitude earthquake did no serious damage to the nuclear plant was the case of the Onagawa nuclear station whose reactors are of the same BWR type as the Fukushima Daiichi station. The Onagawa station is located closest to the epicenter of the earthquake at a distance of 80 versus 150 km for Fukushima Daiichi station. The response spectrum of the earthquake observation record of the site ground partially exceeded the response spectrum of the design-basis seismic ground motion (AESJ, 2014). The IAEA mission in 2012 reported that despite prolonged ground shaking and a significant level of seismic energy input to the plant facilities, the structures, systems, and components (SSCs) of the Onagawa station performed its intended functions without any significant damage (IAEA, 2012).

Second, during the Fukushima Daiichi accident, three reactors went under total station blackout with the loss of off-site and on-site AC power to run pumps initially at about the same time and the loss of DC power to control steam-turbine-driven core cooling systems at different times. Once the reactor went into total station blackout and the steam-turbine-driven core cooling system was no longer available, core cooling must be restored within a relatively short time to prevent severe core damage. As indicated in Figs. 5.1—5.3 with an arrow showing a period of no injection, all the three reactors that melted down experienced a period of no core cooling that was too long before the injection of water from a fire engine was successful. The period of no injection was more than 12 hours for Unit 1, about 9 hours for Unit 2, and 6.5 hours for Unit 3. These periods of no injection needed to be shortened as much as possible to, for example, ~2 hours for Unit 1, ~5 hours for Unit 2, and ~2 hours for Unit 3 in order to keep the core covered with water and avoid severe core damage (NEA, 2015). The purpose of the severe accident countermeasures, as discussed in Section 5, is to make this possible.

3. Investigative studies of lessons learned

In 2012, the US Congress asked the NAS to conduct a thorough technical study on lessons learned from the Fukushima Daiichi nuclear accident for improving safety and security of commercial nuclear power plants in the United States and directed the US Nuclear Regulatory Commission (NRC) to provide funding for the study. A committee of 21 experts covering various aspects of the accident was appointed by the NAS to carry out this study (NAS, 2014). The committee held 39 meetings during the course of this study to gather information for about one and a half year from July 2012 to December 2013. One of the meetings was held in Tokyo for an in-depth discussion with experts in Japan. The NAS report was issued in the second half of 2014, approximately 2 years after the two major investigations, one by the National Diet Independent Investigation Commission (July 5, 2011) and another one by the Cabinet Investigation Committee (July 23, 2011), had released their final reports. So the NAS report had the benefit of digesting information and recommendations available from the prior investigation reports (including the abovementioned two major investigations) in addition to new information available after July 2012.

In addition to the NAS study, there are five other investigative studies that are also worth reading for serious readers and their reports are available online, i.e., the study by the Japan Nuclear Technology Institute (JANTI) (JANTI, 2011), the study by the special committee on Fukushima of the American Nuclear Society (ANS) (ANS, 2012), the study by the TEPCO investigation committee (TEPCO, 2012), the study by the investigation committee of the Atomic Energy Society of Japan (AESJ) (AESJ, 2014), and the study by Professor Ishikawa (2014). The JANTI report was first released in October 2011, followed by the ANS report in March 2102, then the TEPCO investigation report in June 2012, and the AESJ report in September 2013. The AESJ committee started the investigation in August 21, 2012, after the establishment of the Nuclear Regulation Authority (NRA) in June 2012. The TEPCO investigation report is loaded with highly technical contents, such as technical assessments of tsunami inversion, seismic response, and core damage, and examinations of equipment response during the accident. The book by Ishikawa (2014) is presented in an easy-to-follow lecture-note style with in-depth and insightful explanations and discussions of everything that was relevant.

The NAS lesson-learned study report (NAS, 2014) provides an excellent resource for interested readers to dive into the deep details of each aspect of the accident. Here, only a summary of the causes of the Fukushima accident as identified in the NAS study report is provided in Section 3.1, and the summary of lessons learned from the NAS study is provided in Section 3.2. These lessons learned should be useful for improving the safety of commercial nuclear power plants worldwide.

3.1 Causes of the fukushima accident as found by the US National Academy of Sciences study

The NAS lesson-learned study report summarizes the causes of the Fukushima Daiichi accident into the following eight factors (NAS, 2014). These factors were found to prevent plant personnel from achieving success in averting reactor core damage and contributed to the overall severity of the accident.

1. Failure of the plant owner (TEPCO) and the principal regulator (Nuclear and Industrial Safety Agency [NISA]) to protect critical safety equipment at the plant from flooding in spite of mounting evidence that the plant's current design basis for tsunamis was inadequate.
2. The loss of nearly all on-site AC and DC power at the plant—with the consequent loss of real-time information for monitoring critical thermodynamic parameters in reactors, containments, and SFPs and for sensing and actuating critical valves and equipment—greatly narrowed options for responding to the accident.
3. As a result of factors 1 and 2, the Units 1, 2, and 3 reactors were effectively isolated from their ultimate heat sink (the Pacific Ocean) for a period far in excess of the heat capacity of the suppression pools or the coping time of the plant to station blackout.
4. Multiunit interactions complicated the accident response. Unit operators competed for physical resources and the attention and services of staff in the on-site emergency response center.
5. Operators and on-site emergency response center staff lacked adequate procedures and training for accidents involving extended loss of all on-site AC and DC power, particularly procedures and training for managing water levels and pressures in reactors and their containments and hydrogen generated during reactor core degradation.
6. Failures to transmit information and instructions in an accurate and timely manner hindered responses to the accident. These failures resulted partly from the loss of communications systems and the challenging operating environments throughout the plant.
7. The lack of clarity of roles and responsibilities within the on-site emergency response center and between the on-site and headquarters emergency response centers may have contributed to response delays.
8. Staffing levels at the plant were inadequate for managing the accident because of its scope (affecting several reactor units) and long duration.

3.2 Lessons learned from the National Academy of Sciences Study

The NAS lesson-learned study report summarizes the lessons learned from the Fukushima Daiichi accident into the following 6 categories (NAS, 2014).

3.2.1 Lessons learned on new information about hazards

The overarching lesson learned from the Fukushima Daiichi accident is that nuclear plant licensees and their regulators must actively seek out and act on new information about hazards that have the potential to affect the safety of nuclear plants, specifically the following:

1. Licensees and their regulators must continually seek out new scientific information about nuclear plant hazards and methodologies for estimating their magnitudes, frequencies, and potential impacts.
2. Nuclear plant risk assessments must incorporate new information and methodologies as they become available.
3. Plant operators and regulators must take timely actions to implement countermeasures when such new information results in substantial changes to risk profiles at nuclear plants.

3.2.2 Lessons learned on improvement of plant systems, resources, and training

Nuclear plant operators and regulators in the United States and other countries have identified and have been taking useful actions to upgrade nuclear plant systems, operating procedures, and operator training in response to the Fukushima Daiichi accident. In the United States, these actions include the nuclear industry's FLEX (diverse and flexible coping strategies) initiative as well as regulatory changes proposed by the US NRC's Near-Term Task Force (NTTF; see more discussion in Section 5.3).

The nuclear industry and its regulator should give specific attention to improve plant systems in order to enable effective responses to beyond-design-basis events, including, when necessary, developing and implementing ad hoc responses to deal with unanticipated complexities.

3.2.2.1 Plant systems

Attention to availability, reliability, redundancy, and diversity of plant systems and equipment is specifically needed for

- DC power for instrumentation and safety system control;
- tools for estimating real-time plant status during loss of power;
- decay heat removal and reactor depressurization and containment venting systems and protocols;
- instrumentation for monitoring critical thermodynamic parameters in reactors, containments, and SFPs;
- hydrogen monitoring (including monitoring in reactor buildings) and mitigation;

- instrumentation for both on-site and off-site radiation and security monitoring; and
- communications and real-time information systems to support communication and coordination between control rooms and technical support centers (TSCs), between control rooms and the field, and between on-site and off-site support facilities. The quality and completeness of the changes that result from this recommendation should be adequately peer reviewed.

3.2.2.2 Resources and training

In order to improve resource availability and operator training for effective response to beyond-design-basis events, attention is needed to the following:

1. Staffing levels for emergencies involving multiple reactors at a site, that last for extended durations, and/or that involve stranded plant conditions.
2. Strengthening and better integrating emergency procedures, extensive damage mitigation guidelines, and severe accident management guidelines, in particular for
    - coping with the complete loss of AC and DC power for extended periods,
    - depressurizing reactor pressure vessels and venting containments when DC power and installed plant air supplies (i.e., compressed air and gas) are unavailable,
    - injecting low-pressure water when plant power is unavailable,
    - transitioning between reactor pressure vessel depressurization and low-pressure water injection while maintaining sufficient water levels to protect the core from damage,
    - preventing and mitigating the effects of large hydrogen explosions on cooling systems and containments, and
    - maintaining cold shutdown in reactors that are undergoing maintenance outages when critical safety systems have been disabled.
3. Training of operators and plant emergency response organizations, in particular,
    - specific training on the use of ad hoc responses for bringing reactors to safe shutdown during extreme beyond-design-basis events and
    - more general training to reinforce understanding of nuclear plant system design and operation and enhance operators' capabilities for managing emergency situations.

The quality and completeness of the changes that result from this recommendation should be adequately peer reviewed.

3.2.3 Lessons learned on risk assessment of beyond-design-basis events

A "design-basis event" is a postulated event that a nuclear plant system, including its structures and components, must be designed and constructed to withstand without loss of functions necessary to protect public health and safety. An event that is "beyond design basis" has characteristics that could challenge the design of plant structures and components and lead to a loss of critical safety functions. The Great East Japan Earthquake and tsunami were beyond-design-basis events.

Beyond-design-basis events—particularly low-frequency, high-magnitude (i.e., extreme) events—can produce severe accidents at nuclear plants that damage reactor cores and stored spent fuel. Such accidents can result in the generation and combustion of hydrogen within the plant and release of radioactive material to the off-site environment. There is a need to better understand the safety risks that arise from such events and take appropriate countermeasures to reduce them.

3.2.4 Lessons learned on risk concepts for nuclear safety regulations

A DBA is a stylized accident—for example, an LOCA or a transient overpower accident—that is required (by regulation) to be considered in a reactor system's design. The Fukushima Daiichi accident was a beyond-design-basis accident. Other major nuclear accidents (Three Mile Island in 1979 and Chernobyl in 1986) are also considered to be beyond-design-basis accidents.

Four decades of analysis and operating experience have demonstrated that nuclear plant core-damage risks are dominated by beyond-design-basis accidents. Such accidents can arise, for example, from multiple human and equipment failures, violations of operational protocols, and extreme external events. Current approaches for regulating nuclear plant safety, which traditionally have been based on deterministic concepts such as the DBA, are clearly inadequate for preventing core-melt accidents and mitigating their consequences. Modern risk assessment principles are beginning to be applied in nuclear reactor licensing and regulation. The more complete application of these principles in licensing and regulation could help further reduce core-melt risks and their consequences and enhance the overall safety of all nuclear plants, especially the currently operating plants.

3.2.5 Lessons learned on off-site emergency response

Emergency response to the Fukushima Daiichi accident was greatly inhibited by the widespread and severe destruction caused by the March 11, 2011, earthquake and tsunami. Japan is known to be well prepared for natural hazards; however, the earthquake and tsunami caused devastation on a scale beyond what was expected and prepared for. Twenty prefectures on three of Japan's major islands (Hokkaido, Honshu, and Shikoku) were affected by the earthquake and tsunami.

The Fukushima Daiichi accident revealed vulnerabilities in Japan's off-site emergency management. The competing demands of the earthquake and tsunami diminished the available response capacity for the accident. Implementation of existing nuclear emergency plans was overwhelmed by the extreme natural events that affected large regions, producing widespread disruption of communications, electric power, and other critical infrastructure over an extended period. In addition the following were noted:

- Emergency management plans in Japan at the time of the Fukushima Daiichi accident were inadequate to deal with the magnitude of the accident, requiring emergency responders to improvise.

- Decision-making processes by government and industry officials were challenged by the lack of reliable, real-time information on the status of the plant, off-site releases, accident progression, and projected doses to nearby populations.
- Coordination among the central and local governments was hampered by limited and poor communications.
- Protective actions were improvised and uncoordinated, particularly when evacuating vulnerable populations (e.g., the elderly and sick) and providing potassium iodide.
- Different and revised radiation standards and changes in decontamination criteria and policies added to the public's confusion and distrust of the Japanese government.
- Cleanup of contaminated areas and possible resettlement of populations are ongoing efforts 3 years after the accident, with uncertain completion time lines and outcomes.
- Failure to prepare and implement an effective strategy for communication during the emergency contributed to the erosion of trust among the public for Japan's government, regulatory agencies, and the nuclear industry.

3.2.6 Lessons learned on nuclear safety culture

The term "safety culture" is generally understood to encompass a set of attitudes and practices that emphasize safety over competing goals such as production or costs. There is universal acceptance by the nuclear community that safety culture practices need to be adopted by regulatory bodies and other organizations that set nuclear power policies, by senior management of organizations operating nuclear power plants, and by individuals who work in those plants.

Although the Government of Japan acknowledged the need for a strong nuclear safety culture prior to the Fukushima Daiichi accident, TEPCO and its nuclear regulators were deficient in establishing, implementing, and maintaining such a culture. Examinations of the Japanese nuclear regulatory system following the Fukushima Daiichi accident concluded that regulatory agencies were not independent and were subject to regulatory capture (The term "regulatory capture" refers to the processes by which regulated entities manipulate regulators to put their interests ahead of public interests.).

4. The new post-Fukushima regulatory body structure in Japan

Not long after the Fukushima accident, two major public investigations in Japan were undertaken with abundant human and financial resources. One investigation was performed by the National Diet Independent Investigation Commission. The other was performed by the Japan Cabinet Investigation committee. The recommendations of these two investigation reports on the structure of the regulatory body are discussed here.

4.1 Recommendations by the National Diet Independent Investigation Commission

The National Diet of Japan enacted a law creating an Independent Investigation Commission on October 30, 2011, to investigate the Fukushima accident, with the legislative branch's investigative powers to request any necessary documents and evidence. It was the first independent commission created in the history of Japan's constitutional government (NAIIC, 2012). The chairman and nine other members were appointed on December 8, 2011. The investigation was conducted by experts without bias for or against nuclear power. The investigation commission held 19 meetings that were open to public observation and broadcast over the internet (except for the first one) simultaneously in Japanese and English for a period of 6 months from December 19, 2011 to June 9, 2012. The meetings included hearings with selected witnesses to those who held responsible positions at the time of the accident in the government, TEPCO, NISA (a regulator at that time), and Nuclear Safety Commission (NSC) as well as interviews with experts from the United States, France, Ukraine, and Belarus. The commission also held three town hall meetings, visited 12 local municipalities within the designated evacuation area, and conducted interviews and survey of residents and workers at the Fukushima Daiichi accident site. The commission issued its report on July 5, 2012.

The commission concluded in its report, among several other things, that the regulator at that time (i.e., NISA with NSC performing a double-check function) needed to be reformed. The safety of nuclear energy in Japan and the public could not be assured unless the regulators went through an essential transformation process. The entire organization needed to be transformed, not as a formality but in a substantial way. The reform must cover the structure of the electric power industry and the structure of the related government and regulatory agencies as well as the operation processes. The report states that NISA and the operators were aware of the risk of core damage from tsunami prior to March 11, 2011, but no regulations were created and TEPCO did not take any protective steps against such an occurrence. The report made a point that if NISA (who had a negative attitude toward importation of new advances in safety technology from overseas) had passed on to TEPCO the countermeasures that were included in the B.5.b subsection of the US security order that followed the 9/11 terrorist attack in the United States and if TEPCO had put the countermeasures in place, the accident may have been preventable. There were many opportunities for taking preventive measures prior to March 11, 2011, but TEPCO did not take any actions and NISA and NSC went along. The Independent Investigation Commission report (NAIIC, 2012) sent strong messages to Japanese lawmakers and the public at large regarding the regulatory reform needed to prevent this type of accident from happening again in Japan. For this very reason, both NISA and NSC were abolished and the NRA, the new regulatory body, was established in September 2012 (Fig. 5.5).

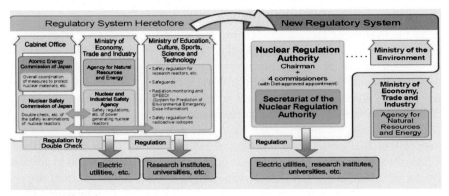

FIGURE 5.5

The birth of Nuclear Regulation Authority in 2012 due to the national regulatory reform in Japan after the March 11, 2011, Fukushima Daiichi accident.

From NRA website (NRA, Nuclear Regulation Authority Leaflet, 2020, http://www.nsr.go.jp/english/e_nra/outline/leaflet.html (accessed 12 February 2020).

4.2 Recommendations by the Government Investigation Committee

Earlier than the appointment of the Independent Investigation Commission by the Diet, the government of Japan by a cabinet decision appointed an investigation committee on the accident at Fukushima power nuclear stations on May 24, 2011, with a mission to make policy recommendations. The committee was composed of 10 members that included scholars, journalists, lawyers, and engineers and was supported by university and government experts. The committee interviewed a large number of people including workers, government officials, and evacuees. The investigation committee published its interim report on December 26, 2011, and its final report on July 23, 2012 (Investigation Committee, 2012).

Similar to the recommendation from the National Diet Independent Investigation Commission, the Cabinet Investigation Committee also recommended, in its report among other things, the establishment of the new organization with regulatory nuclear safety oversight that can function separately from any organization that could unduly influence its decision-making process. The new regulatory body should be given the independence to function and the responsibility to explain nuclear safety issues to the Japanese people while maintaining transparency.

5. Post-Fukushima regulations and technology development

This section discusses the new post-Fukushima safety requirements that have been developed and implemented in Japan and the United States as of February 2020.

5.1 Japan's new regulatory regime

After the Fukushima Daiichi accident in March 2011, there was a sudden loss of confidence in the national nuclear regulatory system among the public in Japan. As a result, an immediate restructuring of the regulatory system was pursued by the government at that time (Fig. 5.5). In June 2012, laws were amended to expand regulatory requirements to cover severe accidents and to allow new requirements to be applied retroactively to existing nuclear power facilities. In September 2012, the NRA was established to replace the NISA, which was the regulatory body at that time (NRA, 2020). NISA was a nuclear regulatory agency under an umbrella of the Ministry of Economy, Trade and Industry (METI) whose mission was to promote the use of nuclear energy. NISA was overseen by the NSC, which was itself responsible for formulating safety policy, and the Atomic Energy Commission (AEC), which is responsible for nuclear power and research policy. The AEC and the now-defunct NSC were both part of the Cabinet Office. NISA's location within METI was seen as compromising its independence and fostering a potential conflict of interest for METI as both promoter and regulator of nuclear energy. NISA was criticized as being too close to both METI and the nuclear industry itself.

NRA is an external organization of the Ministry of Environment with a high degree of independence and with four commissioners selected by the cabinet and approved by votes in both houses. The NRA has enough independence to do its work free from governmental control and undue industry influence. The initial rules (set up by the prime minister at that time) prohibited anyone who had been employed by the nuclear industry in the previous 3 years from being employed by the NRA (WNN, 2014). The rules effectively reduced the pool of suitably qualified candidates for a new organization that needed to grow fast to meet the demand of its services. The NRA initially had a staff of 500 and a budget of ¥50 billion (WNN, 2012). However, the manpower of NRA was strengthened by the merger of NRA and Japan Nuclear Energy Safety Organization (JNES) that was approved by the cabinet on October 25, 2013. By March 2014, the total staff of NRA rose to about 1000 when JNES, a technical support organization with a staff of about 400, merged with NRA. Following its establishment, the NRA had to come up with the new mandatory safety standards that provide countermeasures against natural disasters and major accidents immediately. The new safety requirements document, dated April 2013, has been posted on its website (NRA, 2013a, 2013b). The NRA new safety requirements are directly inspired by lessons learned from the Fukushima Daiichi accident and the difficulties experienced by TEPCO in containing a total station blackout accident brought on by the tsunami flooding (Fig. 5.6). NRA has two immediate and long-term pressing tasks: (1) to process the reactor restart applications that meet the new safety standards and (2) to oversee the decommissioning work at the Fukushima Daiichi site.

After the March 2011 accident at the Fukushima Daiichi plant, all 54 reactors in Japan were gradually taken off line and shut down because of scheduled maintenance outages. Initially, all Japan's nuclear reactors were supposed to undergo stress

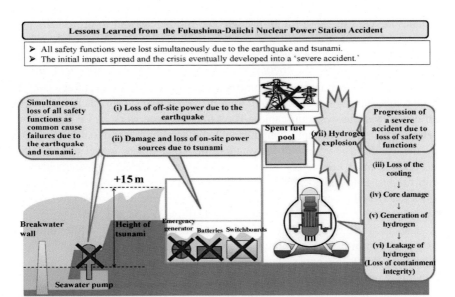

FIGURE 5.6

NRA's perspectives on lessons learned from the Fukushima Daiichi accident.

From NRA website (NRA, Enforcement of the New Regulatory Requirements for Commercial Nuclear Power Reactors, Nuclear Regulation Authority, 2013a. http://www.nsr.go.jp/data/000067212.pdf).

tests at the direction of NISA. However, after the establishment of NRA in July 2013, all reactors are required to be "relicensed" for a restart under the new safety requirements. Only reactors that are relicensed under the more demanding safety requirements would be allowed to restart. Those that are not relicensed would remain shutdown. Under the NRA's reactor restart process, plant operators are required to apply to the NRA in three steps: first, for a permission for "basic design changes" to make changes to the reactor installation; second, for an approval of "detailed design changes" including "construction plan" to strengthen the plant; and third, for approval of the "Revised Tech. Spec." to ensure the unit meets new safety requirements. Operators are required to add certain safety-enhancing equipment including filtered vents and specialized safety facility (backup control room) within 5 years of receiving the NRA's approval of a reactor engineering work program. Examples of equipment installation and construction work include alternate (e.g., mobile) power supplies, multiple sources of cooling water, filtered vents for reducing pressure and removing potentially explosive hydrogen from the PCV in the event of serious core damage, a backup control room in case of loss of the main control room, and a water injection capability to cool a molten core in case of reactor vessel melt-through.

A few plant-specific examples of construction of tsunami protection walls as part of a safety upgrade are as follows. The big construction work at Onagawa Unit 2 includes the construction of a 29-m-high, 800-m-long seawall to protect the plant from potential tsunami at a cost of about ¥340 billion (WNN, 2019a), while at Tokai Unit 2 the work includes a construction of a 17.1-m-high, 1.7-km-long seawall at a cost of ¥180 billion (WNN, 2020a). The safety upgrade work at Mihama Unit 3 includes, for example, improving seismic resistance of the spent fuel storage pool and reactor containment vessel and improving fire protection system such as replacing old cables with fire-resistant cables. The safety upgrade work at Takahama Units 1 and 2 includes, for example, containment building reinforcement (such as reinforcing the existing concrete wall surrounding the containment vessel and installing secondary domes), fire protection system improvement (such as replacing old cables with fire-resistant cables), replacing refueling water storage tank, constructing an anti-tornado wall around the refueling water storage tank, and excavating and constructing a new tunnel for a seawater intake facility (WNN, 2019b).

In addition to the approval for plant restart issued by the NRA, utilities must also get permission from local authorities before the actual plant restart can happen. As of February 2020, only nine reactors have been restarted through this process, and all of them are pressurized water reactors. Only four BWRs (two of which are advanced boiling water reactor [ABWR], while the other two reactors are of the same basic design as Fukushima Daiichi reactors) have been approved by the NRA for a restart, but their prospects of getting a local authority approval have been highly uncertain (Table 5.1). Among all 54 reactors, 21 are older reactors and are being decommissioned. There are 17 more reactors that have submitted their restart application but have not received the restart approval from the NRA.

5.2 New regulatory requirements in Japan

The Fukushima Daiichi accident revealed the weaknesses of the previous regulatory regime. These weaknesses include (1) insufficient design provisions against tsunami, (2) insufficient provisions for accidents far exceeding the postulated design conditions, and (3) unpractical accident management measures under severe accident conditions. According to the NRA document on its website (NRA, 2013a), the new regulatory requirements have been developed by taking into account the lessons learned from the accident at Fukushima Daiichi Nuclear Power Station identified in the reports of the National Diet's Fukushima Nuclear Accident Independent Investigation Commission (NAIIC, 2012) and the Investigation Committee (2012) established by the cabinet considering the harsh natural disaster conditions unique to Japan, which are in line with the safety standards and guidelines of the IAEA (IAEA, 2016, 2019). The new regulatory requirements are much more demanding with an underlying assumption that severe accidents could occur at any moment. The NRA approach for the new regulatory requirements is to evaluate in advance the potential and consequences of a wide spectrum of internal and external accident initiators and to emphasize the importance of the defense-in-depth approach in design and preparations of countermeasures against beyond-design-basis accidents.

The new regulatory requirements aim at primarily (Fukeda, 2014)

Table 5.1 Japanese Reactors that have received Nuclear Regulation Authority restart approval as of February 2020 (JAIF, 2020).

NPP unit no.	Reactor type	Operating company	Date of restart application submitted	Date of restart application approved	Date of restart operation
BWR					
Kashiwazaki-Kariwa 6 and 7	ABWR 1356 MWe	Tokyo Electric	September 2013	Dec. 27, 2017 CP pending	Pending local approval
Tokai 2	BWR 1100 MWe	Japan Atomic	May 2014	Oct. 18, 2018 CP approved	Pending local approval
Onagawa 2	BWR 825 MWe	Tohoku Electric	December 2013	Feb. 26, 2020 CP pending	Pending local approval
PWR					
Sendai 1 and 2	PWR 890 MWe	Kyushu Electric	July 2013	May 27, 2015	Aug. 2015 – Unit 1 Oct. 2015 – Unit 2
Takahama 3 and 4	PWR 870 MWe	Kansai Electric	July 2013	Oct. 10, 2015	Jan. 2016 – Unit 3 Feb. 2016 – Unit 4
Ikata 3	PWR 890 MWe	Shikoku Electric	July 2013	Apr. 19, 2016	Aug. 2016
Ohi 3 and 4	PWR 1180 MWe	Kansai Electric	July 2013	Sep. 2017	Mar. 2018 – Unit 3 May 2018 – Unit 4
Genkai 3 and 4	PWR 1180 MWe	Shikoku Electric	July 2013	Sep. 2107	May 2018 – Unit 3 June 2018 – Unit 4

ABWR, advanced boiling water reactor; BWR, boiling water reactor; CP, construction plan; NPP, nuclear power plant; PWR, pressurized water reactor.

1. changing the definition of DBAs by including prolonged station blackout and loss of ultimate heat sink;
2. enhancing the prevention measures against common cause failures, in particular due to external hazards, by strengthening the diversity/independence;
3. enhancing the prevention of core damage by preparing alternative measures with use of mobile equipment;
4. enhancing the mitigation measures against severe accident to eliminate a large radioactive release from the containment and to minimize the radioactive consequences by mobile and immobile equipment.

The previous pre-Fukushima regulatory requirements did not cover severe accidents. Countermeasures against severe accidents were left purely to the discretion of individual operating companies. The new post-Fukushima regulatory requirements expand the scope to include severe accidents and acts of terrorism such as intentional aircraft crash (Fig. 5.7). The new regulations aim to significantly enhance the design basis and strengthen protective countermeasures against natural phenomena including earthquake, tsunami, volcanic eruption, tornado, and forest fires. The new regulations also aim to enhance countermeasures against events (other than natural phenomena that may trigger common cause failures), such as on-site fire and internal flooding, and reinforce the power supply system to prevent power failure. The new regulations require countermeasures to prevent the spread of severe

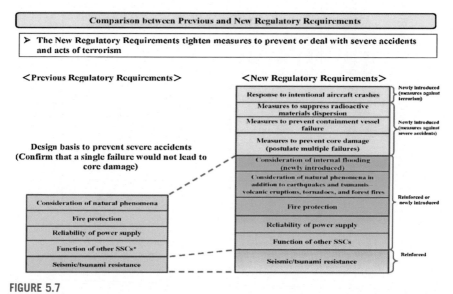

FIGURE 5.7

Japan's new (post-Fukushima) regulatory requirements versus previous (pre-Fukushima) requirements. *SSCs*, structures, systems, and components.

From NRA website (NRA, Enforcement of the New Regulatory Requirements for Commercial Nuclear Power Reactors, Nuclear Regulation Authority, 2013a. http://www.nsr.go.jp/data/000067212.pdf).

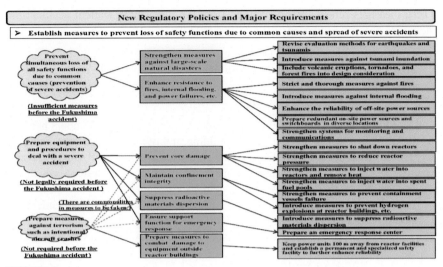

FIGURE 5.8

Japan's new (post-Fukushima) regulatory requirements for severe accident countermeasures.

From NRA website (NRA, Enforcement of the New Regulatory Requirements for Commercial Nuclear Power Reactors, Nuclear Regulation Authority, 2013a. http://www.nsr.go.jp/data/000067212.pdf).

accidents (Fig. 5.8) and encourage the use of both mobile equipment (as used in the FLEX strategy in the United States) and permanent equipment to enhance reliability. Countermeasures against severe accidents must be included in emergency operations. The new regulations requiring countermeasures against intentional aircraft crashes encourage the use of mobile equipment and injection connection points in multiple locations and a specialized safety facility as a backup control room.

5.2.1 Safety upgrade at Kashiwazaki-Kariwa site
TEPCO, the most affected company by the Fukushima Daiichi accident, has taken extra-ordinarily fast steps to safety enhancements of two ABWRs at Kashiwazaki-Kariwa (KK) site. KK Units 6 and 7 are among the first BWRs to submit and receive an approval for their restart application (although the restart operation is pending local authority approval). Here, the safety upgrade at the KK site is used as an example to illustrate the upgrade required to meet the post-Fukushima safety requirements in Japan. The following sections discuss the significant improvements made by TEPCO in power supply and earthquake and tsunami protection as reported by the 2015 IAEA mission to KK Units 6 and 7 (IAEA, 2015).

5.2.1.1 Power supply
The station has implemented comprehensive measures to enhance the AC and DC power systems to supply essential loads required under design extension conditions.

These measures reflect lessons learned from the Fukushima Daiichi accident that include two diverse functions:

- a reliable alternate AC and DC power supply system,
- an alternate mobile unit, containing a spare battery and diesel generator charger, that can be connected to the division I DC system in order to ensure continued DC power for the RCIC system.

In addition, a new, dedicated battery with capacity of at least 72 hours has been installed at high elevation in the reactor building. It provides DC power to dedicated accident instrumentation and SFP water level measurement, as well as power necessary to activate the high-pressure alternative cooling (HPAC) system, which is a newly installed turbine-driven HPAC system.

This combination of reliable AC/DC power systems and an independent and autonomous core injection is an enhancement of the original station design basis. The station would be able to withstand a simultaneous LOCA and a station blackout.

5.2.1.2 Earthquake protection
Significant seismic enhancement modifications have been implemented:

- Enhanced existing SSCs: main exhaust stacks, SFP (sloshing protection), reactor building crane, reactor building roof structure, switching station facility, piping supports.
- New SSCs designed based on the revised design basis earthquake: seawall (tidal embankment), main filter equipment, mobile gas turbine generators, and the TSC that is built on seismically isolated foundations.

5.2.1.3 Tsunami protection
Regarding safety enhancement measures related to tsunamis, a decision was taken by TEPCO after the Fukushima accident to build a protective seawall (tidal embankment) around the safety related area of the station, complemented by supplementary measures such as flood barriers, watertight doors, and waterproofing penetrations.

5.2.1.4 Emergency preparedness
The station maintains a TSC that is organized using the Incident Command System (ICS). Three teams of qualified TSC personnel are in place and 45 individuals are assigned to the planning team, with 9 (a part of ICS functions) required to achieve minimum staffing levels. The team approach is designed to minimize the impact of stress from extended−duration staffing periods.

The TSC is located in a seismically isolated building that has a dedicated heating, ventilation, and air-conditioning (HVAC) system powered by an independent gas turbine generator. The HVAC system maintains the TSC at a positive pressure relative to the outside environment. Lead shielding is provided in aprons hung from windows in the TSC to provide additional radiologic shielding. Although not missile protected, windows in the TSC are covered with a protective coating to prevent shattering.

Facilities, instruments, tools, equipment, and communication systems are maintained through preventive and, when necessary, corrective maintenance programs.

The station maintains a significant amount of portable equipment stored in designated elevated locations. Response to multiunit events is provided by having sufficient equipment to support response at all seven units concurrently. An abundant supply of hoses used for portable makeup to the reactor, containment, and SFP exists for all units at the station. Adequate testing programs are in place for hoses that are maintained in the inventory.

5.3 Post-Fukushima regulatory requirements in the United States

Following the March 11, 2011, accident at the Fukushima Daiichi nuclear power plant, the US NRC established a senior-level agency task force referred to as the NTTF. The NTTF conducted a systematic and methodic review of the NRC regulations and processes to determine whether the agency should make additional improvements in NRC regulations or processes in light of the events at Fukushima Daiichi. The NRC evaluated the lessons learned from the accident and prioritized its recommendations into three tiers based on the urgency of the action, the need for additional information, and the availability of critical skill sets. The most significant of these activities were addressed in tier 1 activities (NRC, 2019c).

As tier 1 activities, the NRC issued the following three orders and one request for information on March 12, 2012:

- EA-12-049, Order with regard to requirements for mitigation strategies for beyond-design-basis external events (NRC, 2012a).
- EA-12-051, Order with regard to reliable spent fuel pool instrumentation (NRC, 2012c).
- EA-12-050, Order with regard to reliable hardened containment vents (NRC, 2012b) (however, this order was later replaced by EA-109 on June 6, 2013 (NRC, 2013)).
- Request for information on seismic and flooding hazard protection, seismic and flooding hazard reevaluations using up-to-date methods, and emergency preparedness communications and staffing capabilities.

As explained in ACRS (2016), on December 6, 2016, the draft final rule 10 CFR 50.155 was recommended to the commission by the Advisory Committee on Reactor Safeguards that addresses requirements for licensees to develop mitigation strategies for beyond-design-basis external events consistently with the intent of NRC Order EA-12-049. The rule relocates and integrates the requirements for licensees to develop and maintain strategies to cope with the loss of a large area of the plant due to explosions or fire. Additionally, the rule incorporates the requirements for licensees to install reliable SFP instrumentation according to NRC Order EA-12-051. By consolidating these requirements, the rule establishes a consistent regulatory basis for licensees to develop coordinated and flexible strategies to maintain or restore core cooling, containment, and spent fuel cooling capabilities for a wide range of conditions that are beyond the licensing design basis for each plant.

The final rule, 10 CFR 50.155, "Mitigation of Beyond-Design-Basis Events," was issued in 2019 (NRC, 2019a) (The final rule did not contain several requirements that were in the draft final rule, such as strategies that apply for the reevaluated seismic and external flooding hazards at each site, enhancements to emergency response capabilities, and staffing and communication capabilities assessment.). This rule requires licensees to develop, implement, and maintain strategies and guidelines to mitigate beyond-design-basis external events. In conjunction with this rule, in June 2019, the NRC issued Regulatory Guide 1.226, "Flexible Mitigation Strategies for Beyond Design Basis Events," (NRC, 2019a) that endorses the methods and procedures promulgated by the Nuclear Energy Institute (NEI) in technical document NEI 12-06, "Diverse and Flexible Coping Strategies (FLEX) Implementation Guide," Revision 4 (NEI, 2016) for meeting in parts the regulations in 10 CFR 50.155. In June 2019, the NRC also issued Regulatory Guide 1.227, "Wide-Range Spent Fuel Pool Level Instrumentation," (NRC, 2019b) that endorses the methods and procedures in the document NEI 12-02, "Industry Guidance for Compliance with NRC Order EA-12-051," Revision 1 (NEI, 2012).

Since December 16, 2011, the NEI had been actively working on proposing FLEX as a strategy to fulfill the safety functions of core cooling, containment integrity, and spent fuel cooling as an industry initiative in response to lessons learned from the Fukushima accident. Over 6 years, the NEI had developed several revisions to NEI 12-06 that were submitted to the NRC, including Revision B (May 4, 2012), Revision 1 (August 25, 2015), Revision 2 (December 10, 2015), Revision 3 (September 22, 2016), and Revision 4 (December 12, 2016), the version that was accepted by the NRC. The guidelines in NEI 12-6, Revision 4 recommend a three-phase approach for mitigating beyond-design-basis external events. The initial phase makes use of installed equipment and resources to maintain or restore key safety functions including core cooling, containment, and SFP cooling. The transition phase includes providing sufficient, portable, on-site equipment and consumables to maintain or restore these functions until they can be accomplished with resources brought from off-site. The final phase includes obtaining sufficient off-site resources to sustain these functions indefinitely.

In terms of implementation, by 2019, all US nuclear plants had completed the implementation of the safety enhancements required by the ier 1 mitigation strategies order EA-12-049 and the SFP instrumentation order EA-12-051 (NRC, 2019c). The NRC had completed the review of the required plans and strategies and on-site verification inspections. The requirements of these two orders were approved by the NRC as a final rule (SECY-16-0142) in the NRC's regulations on January 24, 2019. The final rule was published and has become effective since September 2019.

All applicable operating power reactor licensees have implemented the safety enhancements required by the reliable hardened containment vent order. Verification inspections are ongoing and are expected to continue through 2020 (as of this writing). Because this order applies only to a limited group of plants (i.e., BWRs with Mark I or Mark II containments), the requirements did not need to be codified in the NRC regulations.

All applicable US operating power reactor licensees had completed the seismic and flooding-related inspections and hazard reevaluations for the request for information. Licensees have implemented interim measures, if necessary, while additional evaluation of the impact of the reevaluated hazards on the sites is ongoing. The NRC has reviewed the information provided and has identified those sites where additional evaluations of impact are needed. Some licensees needed to perform a flooding integrated assessment or a seismic probabilistic risk assessment, while others needed to perform limited-scope evaluations. This determination was made based on the degree to which the new flooding or seismic hazard estimates varied from what was assumed during initial licensing. The NRC is using this additional information to determine if additional regulatory actions under the NRC backfit process are warranted. All applicable operating power reactor licensees are expected to have provided the requested information by December 2019 (NRC, 2019c).

5.3.1 Implementation on Gen-III+ reactor

The Westinghouse AP1000 is the latest advanced light water reactor of Gen-III+ being built in the United States at Vogtle nuclear power station near Waynesboro, Georgia. Two units are scheduled to be online by November 2021 for unit 3 and November 2022 for unit 4 (WNN, 2020b). It is interesting to see how the final rule, 10 CFR 50.155, "Mitigation of Beyond-Design-Basis Events," is applied to this type of reactor, the details of which are available in Appendix F of NEI (2016). A summary is provided here. By nature of the passive safety approach and its licensing basis, the AP1000, based on the Design Control Document Rev. 19, is designed to provide a significant coping period for a station blackout. Hence, the focus is on the transition from passive systems operation and their initial coping capabilities (i.e., 72 hours) to indefinite, long-term operation of the passive cooling systems with support using off-site equipment and resources.

For the AP1000 design, the fundamental difference due to the passive nature of the plant design is in the significantly longer coping period available before FLEX equipment may be required (i.e., at least 72 hours) and in the reduced size and number of this equipment. Thus many of the strategies detailed in NEI 12-06 are not required for AP1000. The underlying strategies for coping with extended loss of AC power events involve a three-phase approach. (1) Initial coping is through installed plant equipment, without any AC power or makeup to the ultimate heat sink. For the AP1000, this phase is already covered by the existing licensing basis. This covers the 0- to 72-hour basis for passive systems performance for core, containment, and SFP cooling. (2) Following the 72-hour passive system coping time, support is required to continue passive system cooling. This support can be provided by installed plant ancillary equipment or by off-site equipment installed to connections provided in the AP1000 design. The installed ancillary equipment is capable of supporting passive system cooling from 3 to 7 days. (3) In order to extend the passive system cooling time to beyond 7 days (to an indefinite time), some off-site assistance is required. As a minimum, this would include delivery of diesel fuel oil.

The AP1000 reactor only requires a few small pieces of FLEX equipment. In addition, it is not required for at least 72 hours because of the large passive system coping time. The FLEX equipment required for the AP1000 includes (1) one FLEX self-powered pump (required to provide makeup to the passive containment cooling system (PCS) and SFP) with a capability of 135 gallons per minute (gpm) and 273-ft head, (2) one FLEX electric generator (required to provide postaccident monitoring and emergency lighting) with a capacity of 15 kW and 480 V, and (3) off-site makeup water (only required if on-site makeup water is not available).

For the postaccident monitoring system class 1E instrumentation, the instrumentation is powered for the first 72 hours by the safety-related batteries and is powered thereafter by on-site or off-site diesels for indefinite coping. There are multiple connection points for the FLEX electric generator such that portable instrumentation is not necessary.

5.4 Development of Accident Tolerant Fuel Technology as Risk Mitigation

In addition to the regulatory changes and the implementation of FLEX strategies to effectively reduce the risks and consequences of severe accidents like Fukushima Daiichi, the global nuclear community is also making an effort in developing a new generation of fuel and cladding materials that can survive in severe accidents for longer periods of time than the current zirconium alloy cladding and uranium dioxide (UO_2) fuel (effectively providing more time for operators to respond to an accident). This new generation of fuel and cladding materials have been known as "Accident Tolerant Fuels (ATF)," or Advanced Technology Fuels in certain countries. These ATF technologies have the potential to enhance plant safety by offering better performance during normal operation, transients, and accident conditions. One year after the Fukushima accident, the United State Congress directed the Department of Energy (DOE) to start the R&D of nuclear fuels and claddings with enhanced accident tolerance for light water reactors (LWRs) and provided funding in various stages. Major awards were made to multi-institutional teams led by Westinghouse Electric Company, Global Nuclear Fuel (a GE-led joint venture with Hitachi), and Framatome on a cost share basis, demonstrating the commitment and leadership by the US nuclear industry in developing ATF technologies. DOE also awarded a few universities for teaming up and performing R&D on ATF. The DOE laboratories have been working closely with both industry and academia by establishing critical infrastructures and providing supports to testing and qualification of ATF materials. Under these initiatives, several ATF options have been explored and developed.

These ATF technologies can be made into two main categories: the near-term ATF technologies which have moderate ATF benefits but can be deployed relatively sooner, while the longer-term ATF technologies which offer more potentials in ATF safety benefits but will take longer time in R&D and licensing. The near-term ATF technologies consist of various coated zirconium alloy cladding for enhanced oxidation resistance during the Design Basis Accidents (DBA) and beyond DBA conditions, and doped UO_2 pellets for higher density and higher safety margins in LOCA transients and fuel washouts. The longer-term ATF technologies encompass (1) advanced ceramic composite cladding such as ceramic matrix composite (CMC) cladding and channel boxes (in BWRs) made of silicon carbide fibers in a high purity silicon carbide matrix (SiC_f/SiC_m) for significant improvement in high-temperature corrosion resistance and strength retention, (2) the iron-based cladding such as iron-chromium-aluminum alloy (FeCrAl) for enhanced corrosion resistance beyond DBAs, (3) the high density fuel pellets such as U_3Si_2 and UN with better fuel cycle economics and better thermophysical properties, (4) the refractory metal cladding such as molybdenum, (5) various uranium dioxide (UO_2) pellets with additives for thermal conductivity enhancement, and (6) the fully ceramic microencapsulated (FCM) fuels for ultimate safety improvement (NEA, 2018).

DOE set a development timeline for ATF with the lead test assemblies (LTA) beginning in 2022. Considering that the traditional nuclear fuel development and licensing process takes more than a decade, the schedule mandated by DOE requires a significant push and commitment by the nuclear industry. Yet, the fuel vendors and utilities are ahead of the schedule in ATF testing and commercialization. Fig. 5.9 shows the timeline and key milestones the US nuclear industry has already accomplished and is scheduled to achieve in the future (NEI, 2020).

Global Nuclear Fuel (GNF)'s ATF test products have been loaded into two reactors for testing, Southern Nuclear's Hatch unit 1 and Exelon's Clinton. In March 2018, the Hatch reactor was loaded with lead test assemblies (LTAs) with ARMOR-coated cladding and lead test rods (LTRs) with iron-chromium-aluminum cladding (known as IronClad) in two forms, a rod with no fuel and a solid bar segment. By February 2020, these LTRs completed the first full fuel cycle, and a sampling of these LTRs was transferred from the reactor to the spent fuel pool for inspections. Initial inspections confirmed that the fuel performed as expected. As of this writing, further evaluations of material and coating properties will be performed by Oak Ridge National Laboratory. The Clinton reactor was loaded with IronClad LTAs for testing in late 2019.

In April 2019, Framatome loaded four full-length lead test assemblies containing chromium-coated M5 zirconium alloy cladding and doped UO_2 pellets into Southern Nuclear's Vogtle unit 2 reactor. In September 2019, Westinghouse inserted 20 lead test rods (LTRs) consisting of several EnCore® ATF products into Exelon's Byron nuclear power reactor Unit 2. These ATF products includes Cr-coated zirconium alloy cladding (for enhanced oxidation and corrosion resistance), the higher density ADOPT pellets (chromia and alumina doped UO_2 pellets), and uranium silicide (U_3Si_2) pellets (with higher density and thermal conductivity than standard UO_2 fuel). Test data will help determine the safe operating limits of each cycle and will be used to license the fuels with the NRC. These LTA test programs are at least 3 years ahead of the schedule set by DOE. Both Framatome and Westinghouse are targeted for longer-term ATF technologies by 2022, with the first batch reloads for the near-term ATF products starting as early as 2023 followed by a full core ATF implementation by 2026.

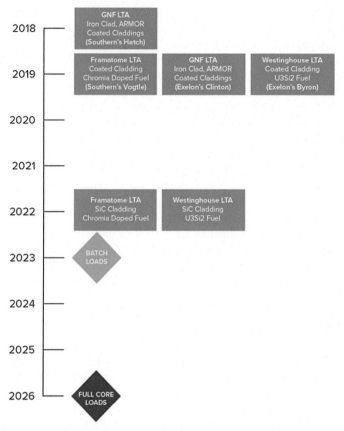

FIGURE 5.9

Key ATF development milestones and schedule for US nuclear industry.

From NEI website 2020 https://www.nei.org/advocacy/make-regulations-smarter/accident-tolerant-fuel.

The ability to timely license ATF is an important part of the commercialization process and has great impact on meeting the schedule. To meet the expedited schedule, the Nuclear Regulatory Commission (NRC) has taken proactive measures to make the ATF licensing processes more efficient and effective. Moreover, the NRC staff recognized the importance of early involvement with the ATF applicants in R&D phase, and has been closed engaged with various stakeholders as part of the effort, including licensees, nuclear fuel vendors, industry groups, nongovernmental organizations, and the NRC's international counterparts.

Similar to USA, several countries started their national programs in ATF R&D, including Japan, Korea, European Union, UK, Russia, and China. Although the focus in technologies may vary from one country to another, and the timeline could also differ, the global expansion of future use of ATF would improve the safety of the current nuclear reactor fleet.

6. Concluding remarks

The post-Fukushima safety regulations require nuclear power plants to be strengthened and equipped with plant systems and equipment and emergency procedures for countermeasures against severe accidents that have or have not occurred in the past. Under the new regulations, post-Fukushima nuclear plants will be able to cope with the complete loss of AC and DC power for extended periods, i.e., the same event that happened to the Fukushima Daiichi nuclear stations. In addition, the post-Fukushima nuclear plants are also figured to be able to cope with other extreme external events (such as a terrorist attack) that have never happened to nuclear plants. Future accidents, if ever happen, will be manageable for nuclear plants under the new safety standards, with no or minimal radiation dose to the public. For example, if a similar extreme earthquake/tsunami event that triggered the Fukushima Daiichi accident is assumed to happen again in a similar way with a loss of off-site power due to damage to power lines, the countermeasures that are already in place in the strengthened nuclear plant will make sure that there are multiple layers of AC and DC power supply to operate the core cooling functions. The upgraded plant modification will not allow the EDGs, batteries, and safety equipment to be flooded. The upgraded tsunami protective seawall will prevent or minimize flooding. The station will very unlikely lose its on-site AC and DC power for survival to a cold shutdown. As long as the on-site AC and DC power is available, the reactor will be kept from overheating and can be safely shut down.

During the Fukushima Daiichi accident, there were 3 other nuclear stations (Onagawa, Fukushima Daini, and Tokai Daini) located on the same east coast of the Honshu island as the Fukushima Daiichi site that were hit by the same tsunami. All operating reactors on these 3 sites were safely shut down because none of the reactors went into a total station blackout. At Onagawa station, the closest to the earthquake epicenter, AC power was available throughout the event. Of the five off-site AC power lines, one was available, whereas the others were tripped by the earthquake. AC power was also available from 6 of 8 EDGs (Kato, 2012) and they supplied power to the plants until off-site power was restored on March 12. At Tokai Daini station, the earthquake cut off all off-site AC power lines, but EDGs were available and supplied power until off-site power was restored on March 13. At Fukushima Daini, not all AC power lines were lost during the earthquake. One AC line continued to supply power, and a second backup line was restored on March 12 and a third on March 13 (NAS, 2014). Most EDGs were rendered inoperable because of the floodwaters, except one in Unit 3. With great efforts from operators, it was fortunate that all these reactors were safely shut down. Most of these reactors were shut down through emergency operating procedures where RCIC and safety relief valves were actuated, whereas a few reactors were shut down through nonemergency procedures (AESJ, 2014). However, in technical terms, they were safe because their design basis that was in place at that time allowed them to survive as long as the event was not a station blackout, in other words, as long as there was AC and DC power. This is what the pre-Fukushima nuclear plants had been designed for. They were protected from meltdown according to their licensing design basis. In a similar way, the post-Fukushima nuclear plants have been strengthened to survive an extended station blackout that pre-Fukushima nuclear plants were not designed for.

Acknowledgment

The author, Wison Luangdilok, would like to thank Isao Kato of Tohoku Electric Company for comments and help in accuracy checking of information not available in major Fukushima investigation reports.

References

ACRS, December 6, 2016. Letter to NRC chairman regarding draft final rule 10 CFR 50.155, "Mitigation of Beyond-Design Basis Events" and Associated Regulatory Guidance, Advisory Committee on Reactor Safeguards (ACRS) (ADAMS Accession No. ML16341B371).

AESJ, 2014. The Fukushima Daiichi Nuclear Accident — Final Report of AESJ Investigation Committee. Springer. https://link.springer.com/book/10.1007/978-4-431-55160-7.

ANS, June 2012. Fukushima Daiichi: ANS Committee Report. American Nuclear Society revised. http://fukushima.ans.org/report/Fukushima_report.pdf.

Fukeda, T., April 8, 2014. Lessons learned from the Fukushima Daiichi accident and responses in new regulatory requirements. In: NRA Presentation to OECD/NEA International Conference on Global Nuclear Safety Enhancement, Tokyo, Japan.

IAEA, July 30—August 11, 2012. IAEA Mission to Onagawa Nuclear Power Station to Examine the Performance of Systems, Structures and Components Following the Great East Japanese Earthquake and Tsunami, International Atomic Energy Agency Mission Report to the Government of Japan, IAEA-2012. https://www.iaea.org/sites/default/files/iaeamissiononagawa.pdf.

IAEA, June 29—July 13, 2015. Report of the Operational Safety Review Team (OSART) Mission to the Units 6 and 7 of Kashiwazaki-Kariwa Nuclear Power Plant. International Atomic Energy Agency, Japan. Report NSNI/OSART/183/2015.

IAEA, 2016. Safety of Nuclear Power Plants: Design, SSR-2/1, Rev. 1. International Atomic Energy Agency, Vienna, Austria.

IAEA, 2019. Accident Management Programs for Nuclear Power Plants, SSG-54. International Atomic Energy Agency, Vienna, Austria.

Investigation Committee, Final Report on the Accident at Fukushima Nuclear Power Stations of Tokyo Electric Power Company, Government of Japan, July 23, 2012. https://www.nirs.org/wp-content/uploads/fukushima/SaishyuRecommendation.pdf, http://www.cas.go.jp/jp/seisaku/icanps/eng/finalgaiyou.pdf.

Ishikawa, M., 2014. A Study of the Fukushima Daiichi Nuclear Accident Process, what Caused the Core Melt and Hydrogen Explosion? Springer. https://link.springer.com/book/10.1007%2F978-4-431-55543-8.

JAIF, 2020. Japan Atomic Industrial Forum. https://www.jaif.or.jp/en/npps-in-japan. (Accessed 16 February 2020).

JANTI, October 2011. Examination of Accident at Tokyo Electric Power Co., Inc.'s Fukushima Daiichi Nuclear Power Station and Proposal of Countermeasures. Japan Nuclear Technology Institute. http://www.gengikyo.jp/english/shokai/Tohoku_Jishin/report.pdf.

Japanese Government, 2011. Report to the IAEA Ministerial Conference on Nuclear Safety, The Accident at TEPCO's Fukushima Nuclear Power Station. https://japan.kantei.go.jp/kan/topics/201106/iaea_houkokusho_e.html.

Kato, I., March 27–28, 2012. Safe Shutdown of the Onagawa Nuclear Power Station—The Closest Boiling Water Reactors to the 3/11/11 Epicenter, Nuclear Symposium, University of Pittsburgh. https://www.thornburghforum.pitt.edu/sites/default/files/Nuclear%20Symposium%20report%20FINAL%20report%2011_5_12.pdf.

Luangdilok, W., 2020. The explosions at Fukushima Daiichi unit 3 and unit 4 and implications on the evaluation of 1F3 accident. Nuclear Engineering and Design 362 (2020), 110536. https://doi.org/10.1016/j.nucengdes.2020.110536.

Mizokami, S., March 6, 2020. Personal Communication Between Wison Luangdilok and Shinya Mizokami of TEPCO Holdings.

NAIIC, The Official Report of the Fukushima Nuclear Accident Independent Investigation Commission, Executive Summary, National Diet of Japan, 2012. https://web.archive.org/web/20120710075620, http://naiic.go.jp/wp-content/uploads/2012/07/NAIIC_report_lo_res.pdf.

NAS, 2014. Lessons Learned from the Fukushima Nuclear Accident for Improving Safety of U.S. Nuclear Plants. The National Academies Press, Washington, DC. https://doi.org/10.17226/18294.

NAS, 2016. Lessons Learned from the Fukushima Nuclear Accident for Improving Safety and Security of U.S. Nuclear Plants: Phase 2. The National Academies Press, Washington, DC. https://doi.org/10.17226/21874.

NEA, 2015. Benchmark Study of the Accident at the Fukushima Daiichi Nuclear Power Plant (BSAF Project) Phase 1 Summary Report. OECD Nuclear Energy Agency. NEA/CSNI/R(2015)18.

NEA, State-of-the-Art Report on Light Water Reactor Accident-Tolerant Fuels, OECD Nuclear Energy Agency, NEA No. 7317, 2018. https://www.oecd-nea.org/science/pubs/2018/7317-accident-tolerant-fuels-2018.pdf.

NEI, August 2012. Industry Guidance for Compliance with NRC Order EA-12-051, 'To Modify Licenses with Regard to Reliable Spent Fuel Pool Instrumentation. NEI 12-20, Revision 1. Nuclear Energy Institute (ADAMS Accession No. ML12240A307).

NEI, September 2016. Diverse and Flexible Coping Strategies (FLEX) Implementation Guide. Revision 4, NEI 12-06. Nuclear Energy Institute (ADAMS Accession No. ML16354B421).

NEI, 2020. Nuclear Energy Institute. https://www.nei.org/advocacy/make-regulations-smarter/accident-tolerant-fuel.

NRA, 2013a. Enforcement of the New Regulatory Requirements for Commercial Nuclear Power Reactors, Nuclear Regulation Authority. http://www.nsr.go.jp/data/000067212.pdf.

NRA, 2013b. Outline of New Regulatory Requirements for Light Water Nuclear Power Plants (Severe Accident Measures), Nuclear Regulation Authority. http://www.nsr.go.jp/data/000067119.pdf.

NRA, 2020. Nuclear Regulation Authority Leaflet. http://www.nsr.go.jp/english/e_nra/outline/leaflet.html. (Accessed 12 February 2020).

NRC, March 12, 2012a. Order EA-12-049, Order Modifying Licenses with Regard to Requirements for Mitigation Strategies for Beyond-Design-Basis External Events. U.S. Nuclear Regulatory Commission. ADAMS Accession No. ML12054A736.

NRC, March 12, 2012b. Order EA-12-050, Issuance of Order to Modify Licenses with Regard to Reliable Hardened Containment Vents. U.S. Nuclear Regulatory Commission (ADAMS Accession No. ML12054A679).

NRC, March 12, 2012c. Order EA-12-051, Issuance of Order to Modify Licenses with Regard to Reliable Spent Fuel Pool Instrumentation. U.S. Nuclear Regulatory Commission (ADAMS Accession No. ML12054A694).

NRC, June 6, 2013. Order EA-13-109, Issuance of Order to Modify Licenses with Regard to Reliable Hardened Containment Vents Capable of Operation under Severe Accident Conditions. U.S. Nuclear Regulatory Commission, Washington, DC (ADAMS Accession No. ML13143A321).

NRC, June 2019a. Regulatory Guide (RG) 1.226, Flexible Mitigation Strategies for Beyond-Design-Basis Events. U.S. Nuclear Regulatory Commission.

NRC, June 2019b. Regulatory Guide (RG) 1.227, "Wide-Range Spent Fuel Pool Level Instrumentation. U.S. Nuclear Regulatory Commission.

NRC, October 2019c. The United States of America Eighth National Report for the Convention on Nuclear Safety, NUREG-1650, Revision 7. U.S. Nuclear Regulatory Commission.

Rempe, J. (Ed.), August 2019. US Efforts in Support of Examinations at Fukushima Daiichi 2019 Evaluation, US Department of Energy. ANL/LWRS-19/08. https://publications.anl.gov/anlpubs/2019/09/154944.pdf.

Sehgal, B.R. (Ed.), 2012. Nuclear Safety in Light Water Reactors: Severe Accident Phenomenology. Elsevier Press.

TEPCO, June 20, 2012. The Fukushima Nuclear Accidents Investigation Report. http://www.tepco.co.jp/en/press/corp-com/release/2012/1205638_1870.html.

TEPCO, December 13, 2013. First Progress Report: The Results of the Investigation and Examining on the Unidentified and Unsolved Matters of the Fukushima Nuclear Accident. https://www4.tepco.co.jp/en/press/corp-com/release/2013/1233101_5130.html.

TEPCO, December 17, 2015. The 4th Progress Report on the Investigation and Examination of Unconfirmed and Unresolved Issues on the Development Mechanism of the Fukushima Daiichi Nuclear Accident. https://www4.tepco.co.jp/en/press/corp-com/release/2015/1264602_6844.html.

TEPCO, December 25, 2017. The 5th Progress Report on the Investigation and Examination of Unconfirmed and Unresolved Issues on the Development Mechanism of the Fukushima Daiichi Nuclear Accident, Attachment 3—10. https://www7.tepco.co.jp/newsroom/press/archives/2017/1485273_10469.html.

WNN, 2012. World Nuclear News. https://www.world-nuclear-news.org/Articles/New-Japanese-regulatory-regime-as-restarts-approac.

WNN, 2014. World Nuclear News. https://www.world-nuclear-news.org/Articles/First-reshuffle-at-Japanese-regulator.

WNN, 2019a. World Nuclear News. https://www.world-nuclear-news.org/Articles/Regulatory-approval-for-Onagawa-2-restart.

WNN, 2019b. World Nuclear News. https://www.world-nuclear-news.org/Articles/Kansai-reschedules-restarts-of-three-reactors.

WNN, 2020a. World Nuclear News. https://www.world-nuclear-news.org/Articles/Safety-upgrades-impact-operation-of-Japanese-units.

WNN, 2020b. World Nuclear News. https://www.world-nuclear-news.org/Articles/Containment-concreting-milestone-at-Vogtle-3.

Nuclear off-site emergency preparedness and response: key concepts and international normative principles*

Günther Handl

Eberhard Deutsch Professor of Public International Law, Tulane University Law School, New Orleans, LA, United States

1. Introduction

As the 10-year anniversary of the Fukushima Daiichi nuclear power plant accident approaches, an important and legitimate question is whether and to what extent its key lessons implicating nuclear safety have been learned. Clearly, severe accident[1] management practices at nuclear installations generally, and—considering the Fukushima accident—triggering combined effects of earthquake and tsunami[2]—the mitigation of so-called "beyond-design-basis events" specifically,[3] have garnered sustained attention by regulators, the industry and the public.[4] These

---

* This chapter is a revised and updated version of a paper previously published as "Nuclear Off-Site Emergency Preparedness and Response: Some International Legal Aspects," in J.L. Black-Branch and D. Fleck (Eds.), Nuclear Non-Proliferation in International Law-Volume III: Legal Aspects of the Use of Nuclear Energy for Peaceful Purposes (Springer/Asser Press, The Hague, 2016) 311–354.

[1] That is an accident beyond the nuclear power plant's design basis. A design basis accident or "maximum credible accident" involves "accident conditions against which a facility is designed according to established design criteria, and for which the damage to the fuel and the release of radioactive material are kept within authorized limits." See IAEA (2009).

[2] See, The Fukushima Daiichi Accident (2015): "The vulnerability of the Fukushima Daiichi nuclear power plant to external hazards had not been reassessed in a systematic and comprehensive manner during its lifetime. At the time of the accident, there were no regulatory requirements in Japan for such reassessments, and relevant domestic and international operating experience was not adequately considered in the existing regulations and guidelines. The regulatory guidelines in Japan on methods for dealing with the effects of events associated with earthquakes, such as tsunamis, were generic and brief, and did not provide specific criteria or detailed guidance."

[3] Note, e.g., the discussion in the United States: Nuclear Regulatory Commission (2019).

[4] See, e.g., The Fukushima Daiichi Accident, *supra* note 2; U.S. Nuclear Regulatory Commission (2011); International Experts' (2014); Institute of Nuclear Power Operations (2012); Western European Nuclear Regulators' Association (WENRA) (2014); and National Research Council (2014).

Advanced Security and Safeguarding in the Nuclear Power Industry. https://doi.org/10.1016/B978-0-12-818256-7.00006-4

efforts have aimed at strengthening mitigation capabilities both on-site[5] and at pre-venting serious, long-term radiological effects off-site.[6] In fact, under the impact of Fukushima, avoidance of significant off-site effects has become a critical safety objective, indeed a mantra for both nuclear regulators and the industry.[7] Thus, in February 2015, a specially convened meeting of the Contracting Parties to the Convention on Nuclear Safety (CNS) adopted the Vienna Declaration on Nuclear Safety (Vienna Declaration on Nuclear Safety, 2015). This statement acknowledges as a key objective in implementing the Convention the mitigation of "possible releases of radionuclides causing long-term offsite contamination and … [avoidance of] early radioactive releases or radioactive releases large enough to require long-term protective measures and actions (Vienna Declaration on Nuclear Safety, 2015, Paragraph 2)." Avoidance of off-site contamination is similarly endorsed in the European Council Directive 2014/87/EURATOM[8] as an EU-wide safety objective legally binding upon EU member states. It is, of course, also a key objective in the development of new—Generation IV—"inherently safe" nuclear energy systems (Gen IV International Forum, 2013).

Safety in the design, construction, and operation of nuclear fuel cycle facilities is premised on a defense-in-depth philosophy.[9] It aims at the prevention or mitigation of the effects of a malfunction or accident involving the release of radioactive materials through multiple independent and redundant layers of protection to compensate for potential human and mechanical failures so that no single layer, no matter how robust, is exclusively relied upon (For further discussion, see IAEA (1996)).

---

[5] Thus the Nuclear Safety Review 2014 concludes that the "nuclear industry needs to continue focusing resources on improving severe accident management capabilities because this capability is the key to the success of defense in depth level 4—the last line of defence prior to the on-set of significant off-site consequences." IAEA (2014). See also the presentations and discussions at the IAEA Experts' (2014); IAEA (2015).

[6] See, e.g., IAEA (2012): "The displacement of people and the land contamination after the Fukushima Daiichi accident calls for all national regulators to identify provisions to prevent and mitigate the potential for severe accidents with off-site consequences."

[7] For example, in 2014, Switzerland proposed an amendment to Article 18 of the Convention on Nuclear Safety (CNS). The proposal sought to establish the objective of preventing serious off-site effects as a legal obligation applicable to the design, construction, and operation of nuclear power plants, both new and already existing. However, the Swiss proposal did not win approval at the specially convened diplomatic conference of the Contracting Parties to the CNS in February 2015.

[8] See Article 8 (a) of the European Council Directive 2014/87/EURATOM of July 8, 2014, amending Directive 2009/71/EURATOM establishing a Community framework for the nuclear safety of nuclear installations [2014] OJ L 219/42 [hereafter: "Safety Directive"].

[9] See, e.g., CNS, Article 18 (i) that requires installation states to ensure that "the design and construction of a nuclear installation provides for several reliable levels and methods of protection (defense in depth) against the release of radioactive materials." Thus "[a] key to a defense-in-depth approach is creating multiple independent and redundant layers of defense to compensate for potential failures and external hazards so that no single layer is exclusively relied on to protect the public and the environment." See U.S. Nuclear Regulatory Commission, *supra* note 4, at 11.

In this scheme of things, prevention of abnormal operational occurrences and system failures is the primary or first-level objective, whereas mitigation of the radiological consequences of significant external releases through offsite emergency response measures is a last, albeit exceedingly important goal.[10] Since for the foreseeable future—perhaps until so-called "inherently safe nuclear reactors"[11] become a reality—major accidents entailing off-site releases of radioactive materials cannot be ruled out, EPR covering off-site radiological consequences remains a salient feature of any nuclear safety regime.[12] This essential point was driven home dramatically by the Fukushima accident,[13] much like the disaster involving the Chernobyl nuclear power plant (Executive Summary 3—5; and NEA/OECD, 2002) earlier. At the same time, Fukushima revealed all too clearly the urgent need to revisit and improve off-site emergency preparedness and response (EPR) capabilities globally.[14]

In response, the 2011 IAEA Action Plan on Nuclear Safety called upon member states and other stakeholders to strengthen nuclear safety through a wide range of measures including specific steps regarding EPR (IAEA Action Plan on Nuclear Safety). This call was taken up the following year at the 2nd Extraordinary Meeting of the Contracting Parties to the CNS.[15] Since then, the goal of improving nuclear EPR has been on the agenda of numerous other fora as well. Today, there is no denying that EPR reviews at national, regional, and global levels have deepened awareness of EPR's critical function in nuclear accident management, generated specific policy or institutional changes,[16] and led to notable regulatory action,

---

[10] The last objective of the defense-in-depth approach to nuclear safety "is mitigation of the radiological consequences of significant external releases through the offsite emergency response." See, e.g., Safety Directive, *supra* note 8, at preamble, Paragraph 17.

[11] However, even such reactors are likely to continue to pose some risk. See, e.g., IAEA (2017). This assessment contrasts, of course, with ambitious safety and reliability goals that drive the development of next generation or "Generation IV" nuclear energy systems, which postulate the elimination of "the need for offsite emergency response." See, Gen IV International Forum (2013).

[12] See, e.g., the conclusions of NEA/CNRA/CSNI (2014): "Recognising that all levels of DiD are important in providing adequate protection to the public and enhancing nuclear safety including Level 4 mitigation and Level 5 protective measures (Off-site emergency response and accident management) set down for offsite release."

[13] See, e.g., The Fukushima Daiichi Accident, *supra* note 4, at 74—93 (2015); OECD/NEA (2016).

[14] Indeed, many the shortcomings of Japan's EPR performance following the accident at Fukushima are not the exception to the rule, but rather symptomatic of deficiencies elsewhere on the part of the industry and regulatory authorities generally. See NEA/CNRA/CSNI Joint Workshop, *supra* note 12, at 13; and The Fukushima Daiichi Accident, *supra note* 4, at 7—15 and 96—99. For further details, see IAEA (2015).

[15] Final Summary Report, 2nd Extraordinary Meeting of the Contracting Parties, *supra* note 6, at Paragraph 21 (offering a similarly detailed list of EPR-related measures that warranted special attention).

[16] See, generally, the relevant summary in Progress in the Implementation of the IAEA Action Plan on Nuclear Safety, Report by the Director General, IAEA Doc. GOV/INF/2015/13-GC(59)/INF/5, July 31, 2015, Paragraph 47. As regards institutional innovation, note, for example, the emergence of an Emergency Preparedness and Response Standards Committee (EPReSC) within the Agency's Commission on Safety Standards. See ibid. at 2.

including the revision of IAEA Safety Requirements governing EPR (IAEA, 2015). Nevertheless, it is widely acknowledged that the robustness of national nuclear EPR arrangements remains a matter of concern.[17] At the same time, as some studies question the advisability of emergency evacuations (Thomas, 2017) following the Chernobyl and Fukushima accidents (The Guardian, 2018; The Times, 2017), and the deployment of nuclear power systems with enhanced safety features promises greater safety margins,[18] some post-Fukushima assumptions about EPR policies and regulations are coming under new scrutiny.

It is against the background of these contending claims and developments that this chapter reviews basic principles of nuclear off-site EPR. By necessity, it does so by focusing exclusively on nuclear EPR that aims directly at mitigating off-site effects. A discussion of all relevant EPR aspects—whether directly or indirectly affecting the nature and extent of off-site radiological effects—would simply be beyond the scope of this chapter.[19] For similar reasons of economy, this chapter addresses only off-site EPR during the early (or emergency) and intermediate phases of a nuclear accident[20] and thus will not specifically examine EPR as it relates to the so-called "recovery phase."[21] Finally, this review approaches the issues from the

---

[17] See, e.g., IAEA (2016) , noting the "future challenges" of ensuring that national EPR measures are "robust, resilient, and adequate"; See also ENSREG (2015), noting that although "improvements in emergency preparedness and response had been made since the previous ENSREG conference … the question of whether enough had been done remained"; and Nuclear Transparency Watch(2015).

[18] Unlike the stricken reactors at Fukushima—generation II boiling water reactors—newer, generation III reactors incorporate passive safety features and rely on gravity and natural convection. They are resistant to high temperatures and arguably aircraft crashes. In other words, they offer a much-improved risk profile. See, e.g., Tomić(2018), ESI-CIL Nuclear Governance Project (paper on file with author). See generally Goldberg and Rosner (2011).

[19] Whether and to what extent a nuclear power plant accident will entail significant radiological off-site effects is a function of a huge number of variables associated with both off-site and on-site EPR. The latter includes, notably, the protection of on-site emergency workers.

[20] See IAEA (2007): "The period of time from the detection of conditions warranting an *emergency response* until the completion of all the actions taken in anticipation of or in response to the radiological conditions expected in the first few months of the emergency. This phase typically ends when the situation is *under control*, the *off-site* radiological conditions have been characterized sufficiently well to identify where food restrictions and *temporary relocation* are *required*, and all *required* food restrictions and *temporary relocations* have been implemented." "During these phases the source and releases from the plant have been brought under control. Also, environmental measurements of radioactivity and dose models are available to project doses to members of the public and base decisions on additional protective actions such as food and water interdictions." National Research Council, Lessons Learned, *supra* note 4, at 197.

[21] "Post-accident recovery includes: the remediation of areas affected by the accident; the stabilization of damaged on-site facilities and preparations for decommissioning; the management of contaminated material and radioactive waste arising from these activities; and community revitalization and stakeholder engagement." See The Fukushima Daiichi Accident, supra note 4, at 15, footnote 16.

perspective of international law and public policy. Admittedly, EPR, like the regulation of nuclear safety generally, is in principle a national responsibility.[22] However, national EPR measures and policies can and, more often than not, do have intrinsically international implications. This is obviously so whenever an accident in a nuclear installation state might cause serious transboundary radiological effects and thus necessitates emergency preparedness measures to protect the public and the environment in areas beyond the installation state's jurisdiction and control. Additionally, national EPR measures become a matter of international concern if they are capable of undermining public confidence in nuclear safety practices of other states, regionally or globally.[23] In short, international normative expectations bear on and shape domestic EPR policies, plans, and activities. Indeed, international public policy and law represent the fundamental conceptual framework that, it is safe to say, underpins national nuclear off-site EPR worldwide.

2. The international normative setting for emergency preparedness and response

2.1 The IAEA-centered regulatory framework

As noted, whenever a state is forced to adopt protective measures for the sake of its population and environment against the risk of nuclear contamination emanating from a nuclear installation across the border, off-site EPR in respect of that installation is a matter of intrinsic international concern. Such a situation might raise fundamental questions about present-day limitations of the sovereignty of the state over its own territory. However, there is no need here to examine in detail the basic principles and rules of customary international law applicable to such a situation.[24] For insofar as EPR-related aspects of the relationship between the installation state and risk-exposed neighboring state(s) are concerned, generic customary legal rules

---

[22] See, generally, IAEA Action Plan, at 1: "[I]t is important to note that … [t]he responsibility for ensuring the application of the highest standards of nuclear safety and for providing a timely, transparent and adequate response to nuclear emergencies … lies with each Member State and operating organization." See also Foreword by Yukiya Amano, Director General, in IAEA, IAEA (2015).

[23] Several scenarios might give rise to such a situation as, for example, when a state's EPR measures are clearly at odds with neighboring countries' EPR approaches or inconsistent with applicable international standards. For further discussion, see infra text at notes 126–133. A national EPR program would be similarly problematic if an installation state suffers from what might be called an "embarking country problem"—worrisome deficiencies in regulatory competences, including in the field of EPR, either on account of insufficient resources for, or inattention to, the requisite regulatory infrastructure. As to the continued existence of this problem, see Nuclear Safety Review 2014, *supra* note 5, at 43.

[24] See instead Handl, "Preventing Transboundary Nuclear Pollution: A Post-Fukushima Legal Perspective," in Jayakumar et al. (2015).

have largely been subsumed in, refined by, and expanded upon by specific nuclear conventional law as well as other normative elements.[25] It is thus this international legal setting—a mixture of conventional law, formally nonbinding safety standards and institutional practices shaping normative expectations regarding EPR—that is the focus of this chapter.[26]

2.1.1 Relevant international treaty instruments

The two multilateral instruments most relevant in the present context—because cornerstones of the international EPR framework—are the Convention on Early Notification of a Nuclear Accident[27] and the Convention on Assistance in the Case of a Nuclear Accident or Radiological Emergency,[28] both adopted in the immediate aftermath of the Chernobyl accident.[29] The Convention on Early Notification applies to nuclear accidents that have the potential for an "international transboundary release that could be of radiological safety significance for another State."[30] It requires states parties to notify affected states—directly or through the IAEA, and the IAEA itself—of any nuclear accident involving specifically enumerated facilities and activities (Article 2). The installation state's report must include data regarding inter alia the accident's time, location, radiation releases, and other data essential for assessing the situation, as well as information on off-site protective measures taken or planned (Article 5). Although the Convention on Early Notification itself accords the IAEA an important, though seemingly modest role as an information clearing-house (Article 4), post-Fukushima the Agency has turned into an indispensable manager of the international nuclear emergency information system.[31] By contrast, the Assistance Convention itself assigns to the Agency an evidently central function in the management of nuclear emergency assistance.[32] Against the background of states parties' general obligation to cooperate between themselves and with the Agency to facilitate prompt assistance (Article 1, Paragraph 1), the Assistance

---

[25] For a summary of EPR-related features of the IAEA-centered international legal framework, see also Rautenbach, Tonhauser, and Wetherall, "Overview of the International Legal Framework Governing the Safe and Peaceful Uses of Nuclear Energy – Some Practical Steps," in OECD/NEA (2006).

[26] Some related international normative principles and standards bearing on human rights or procedural entitlements, such as the so-called Waseda Recommendations on human rights and medical management in nuclear disasters, or the UN ECE Convention on Access to Information, Public Participation in Decision-making and Access to Justice in Environmental Matters (Aarhus Convention), are part of this normative matrix also. However, they will not be specifically discussed unless they are of direct and major significance to the issues under review.

[27] IAEA Doc. INFCIRC/335, November 18, 1986.

[28] IAEA Doc. INFCIRC/336, November 18, 1986.

[29] For a discussion, see, e.g., Moser, "The IAEA Conventions on Early Notification of a Nuclear Accident and on Assistance in the Case of a Nuclear Accident or Radiological Emergency," in OECD/NEA (2006), *supra* note 25, at 119; and Adede (1987).

[30] Article 1, Paragraph 1. For further discussion of the notification threshold issue, see *infra* text at notes 183–188.

[31] For details, see *infra* text at notes 196–208.

[32] Convention on Assistance, *supra* note 28.

Convention establishes the Agency as a channel for communication for requests for assistance (Article 2, Paragraphs 1 and 3) and as the repository of individual state-supplied information on national "experts, equipment and materials which could be made available for the provision of assistance to other States Parties … as well as the terms … under which such assistance could be provided" (Article 2, Paragraph 4). Additionally, the Agency is to receive, from each state party, information on national competent authorities and points of contacts (Article 4), critical "go-to addresses" in a nuclear emergency. Finally, the Agency is entrusted with the task of promoting, facilitating, and supporting cooperation between states (Article 1, Paragraph 3). For that purpose, when so requested, it is specifically called upon to collect and disseminate to states EPR-related information (Article 5, Paragraph [a]) to provide assistance with, inter alia, the preparation of emergency plans, the training of emergency personnel, the development of radiation monitoring programs and procedures, and the initial assessment of the accident or emergency (Article 5, Paragraphs [b and c]).

Beyond the Early Notification and Assistance Conventions, special mention must be made of the two nuclear safety conventions, i.e., the Convention on Nuclear Safety[33] and the Joint Convention on the Safety of Spent Fuel Management and on the Safety of Radioactive Waste Management.[34] Indeed, these two conventions play an extraordinarily important role as regards off-site EPR. Apart from a generic safety obligation bearing on EPR in Article 1,[35] the CNS specifically addresses emergency preparedness in Article 16. It requires an installation state to ensure that off-site emergency plans are in place and routinely tested and that its own population as well as "the competent authorities of … [neighboring] states in the vicinity of the nuclear installation are provided with appropriate information for emergency planning and response."[36] Conversely, contracting parties, which do not have a nuclear installation on their territory but are likely to be affected in the event of a radiological emergency at an installation in a neighboring state, are required to adopt and test emergency plans for their own territory.[37] Analogous provisions can be found in the Joint Convention, namely in its Articles 1[38] and 25.[39]

---

[33] IAEA Doc. INFCIRC/449, July 5, 1994.

[34] IAEA Doc. INFCIRC/546, December 24, 1997.

[35] Thus, Article 1, Paragraph (iii) lists among the objectives of the Convention the prevention of accidents with radiological consequences and to mitigate such consequences should they occur.

[36] CNS, Article 16, Paragraphs 1 and 2, respectively.

[37] Article 16, Paragraph 3.

[38] Its Paragraph (iii) lists among the Convention's objectives "to mitigate [accident] … consequences should they occur during any stage of spent fuel or radioactive waste management."

[39] Its Paragraph 1 requires off-site emergency plans, as "appropriate" before and during operation of a spent fuel or radioactive waste management facility as well as the testing of such plans at an appropriate frequency. Paragraph 2 establishes similar obligations for any country likely to be affected in the event of a radiological emergency at a spent fuel or radioactive waste management facility "in the vicinity of its territory."

2.1.2 EPR-related IAEA safety standards, operational arrangements, and services

A second category of parameters of global normative significance for EPR are IAEA safety standards (safety fundamentals, safety requirements, and safety guides[40]) and technical guidance documents.[41] The IAEA is authorized to establish or adopt "standards of safety for protection of health, and the minimization of danger to life and property" against ionizing radiation.[42] And in respect of EPR it has done so—jointly with other international organizations and bodies—by publishing a number of safety requirements and safety guides.[43] EPR safety standards in turn are supported by a series of technical guidance and tools documents covering specific aspects of EPR (examples include generic assessment procedures for determining protective actions during a reactor accident and generic procedures for medical response during a nuclear or radiological emergency).[44]

Given the complexity of EPR-related conventional and Agency safety standards and the resulting challenge for states to implement or abide by them domestically, there is an obvious need for international institutional assistance. As the Agency notes, "[t]he practical implementation of the various articles of the … [Early Notification and Assistance] Conventions as well as fulfillment of certain obligations under Article 16 of the … [CNS] and Article 25 of the … [Joint Convention], warrant the establishment of appropriate arrangements for emergency preparedness and response."[45] In line with this, the Agency's Operations Manual for Incident and Emergency Communication (IAEA, 2020) provides guidance to member states, states parties, and relevant international organizations on their interaction with each other and with the IAEA Secretariat within the framework of the Early Notification and Assistance Conventions (IAEA, 2020 at 4). Further, the Agency acts as the main coordinating body for the Joint Radiation Emergency Management Plan of the International Organizations (JPLAN, 2017). The JPLAN is intended "to support and underpin the efforts of national governments and ensures a coordinated and harmonized international response to radiation incidents and emergencies" among

---

[40] See IAEA(2016): "The Safety Fundamentals … present the fundamental safety objective and principles of protection and safety and provide the basis for the safety requirements"; "Safety Requirements establish the requirements that must be met to ensure the protection of people and the environment, both now and in the future. The requirements are governed by the objective and principles of the Safety Fundamentals." "Safety Guides provide recommendations and guidance on how to comply with the safety requirements, indicating an international consensus that it is necessary to take the measures recommended (or equivalent alternative measures)."

[41] For more details, see IAEA, The Fukushima Daiichi Accident, *supra* note 14, at 123–126.

[42] IAEA Statute, Article III, Paragraph 6.

[43] They are GSR Part 7 *supra* note 23; IAEA (2007); IAEA (2011); and IAEA (2014). Recognizing the importance of the cross-cutting nature of its EPR work, the Agency also established an Emergency Preparedness and Response Standards Committee (EPReSC). See IAEA (2015), Paragraph 31.

[44] For further details, see http://www-ns.iaea.org/tech-areas/emergency/technicalproducts.asp?s=1, visited January 12, 2020.

[45] IAEA, (2015), at preface.

relevant international organizations (JPLN (2017). at v). Additionally, to strengthen the implementation of the Assistance Convention, the Agency facilitates international emergency response through a network of "teams suitably qualified to respond to nuclear or radiological emergencies rapidly and, in principle, on a regional basis (IAEA, 2018). The latest incarnation of this effort is the response and assistance network (RANET), a system for providing international assistance, upon request from a state, following a nuclear or radiological incident or emergency.[46]

At the global level, then, the EPR conventions–based international system for dealing with nuclear or radiological events relies heavily on the central role of the IAEA Secretariat in the coordination of the flow of information and assistance.[47] To better be able to exercise these functions and meet its obligations under the conventions, in 2005 the Agency established an Incident and Emergency Center (IEC) as the global focal point for EPR for nuclear and radiological safety or security related incidents, emergencies, threats, or events of media interest.[48] The Agency plays also an important role in the assessment and evaluation of states' EPR capabilities, both prospectively and during an actual nuclear emergency or incident. To begin with, in 2015, the IEC launched the Emergency Preparedness and Response Information Management System (EPRIMS), an interactive, web-based self-assessment tool that allows member states to share information with other select member states and discuss national arrangements for EPR.[49] Significantly, "EPRIMS … is able to identify where the response arrangements are consistent with IAEA Safety Standards and where further improvement is necessary (Meschen-moser, 2015). Secondly, the IAEA provides two peer review services that are of special interest here, namely the Emergency Preparedness Review (EPREV) Service, which assesses a country's EPR capabilities against current international safety standards and good practices (IAEA, 2018), and the Integrated Regulatory Review Service (IRRS), which assesses the effectiveness of a state's regulatory framework for the safety of its nuclear installations principally against IAEA safety standards.[50] Thirdly, to test international EPR capabilities, the Agency, jointly with the Inter-Agency Committee on Radiological and Nuclear Emergencies (IACRNE), prepares and conducts periodic "convention exercises" (ConvEx exercises)[51] ranging from

---

[46] See IAEA (2018) at 9. For further discussion, see *infra* text at notes 241–249.

[47] For a detailed discussion of the management of emergency information, see *infra* text at notes 196–208. It might be noted also that IACRNE provides a similar coordinating mechanism for those international organizations whose functions or responsibilities include EPR as well.

[48] IEC functions focus on EPR-related safety standards; appraisal services; EPR capacity building; the interagency EPR framework; and emergency assistance. See IAEA (2020).

[49] Report of the Seventh Meeting of Representatives of Competent Authorities identified under the Convention on Early Notification of a Nuclear Accident and the Convention on Assistance in the Case of a Nuclear Accident or Radiological Emergency, Doc. CAM/REP/2014, TM-45,386, July 24, 2014, 10, Paragraph 56.

[50] See IAEA (2013), foreword and 7–9.

[51] See Operations Manual for Incident and Emergency Communication (IAEA, 2020), at 37–44.

the testing of emergency communication links with contact points in member states,[52] of specific parts of the international response system,[53] all the way to full-scale exercises over several days covering severe incidents.[54] Finally, in the wake of Fukushima, the IAEA Secretariat exercises an expanded role in "accident assessment and prognosis,"[55] an important responsibility that is more fully discussed in the following.[56]

2.1.3 The "normative pull" of emergency preparedness and response—related safety standards

Although, generally speaking, IAEA safety standards are not legally binding in a formal sense and instead are merely of a recommendatory nature, they nevertheless exert highly significant normative effects. Firstly, they are binding upon the Agency itself in relation to its own operations and on states in relation to operations assisted by the IAEA, such as the EPREV or IRRS services. Secondly, international organizations that have sponsored a safety standard—in the case of the 2015 EPR safety requirements a total of 13[57]—are expected to apply the standards concerned in their own operations in line with their mandates.[58] Thirdly, given that IAEA safety standards reflect regulatory and industry best practices, they carry authoritative weight.[59] Fourthly, and more importantly, the Agency's safety standards acquire de facto normativity as a result of the way in which obligations under the two safety conventions are to be interpreted and implemented, a fact that continues to be underappreciated by some commentators (Durand-Poudret, 2015). As is well known, the safety conventions incorporate by reference internationally formulated or endorsed standards and criteria, as either capable of providing "guidance on contemporary means of achieving a high level of safety" or of informing the state's obligation

---

[52] So-called ConvEx-1 exercises.

[53] So-called ConvEx-2 exercises.

[54] ConvEx-3 exercises.

[55] The 2011 Action Plan on Nuclear Safety calls upon the Secretariat "… to provide Member States, international organizations and the general public with timely, clear, factually correct, objective and easily understandable information during a nuclear emergency on its potential consequences, *including analysis of available information and prognosis* of possible scenarios based on evidence, scientific knowledge and the capabilities of Member States"." Ibid. at 6 (emphasis added). Obviously, "analysis and prognosis" is a function that is antecedent to, thus distinguishable from, the communication of relevant accident information. See also IAEA(2015).

[56] See *infra* text at notes 202–208.

[57] See GSR Part 7, Preface.

[58] Ibid.

[59] Indeed, it is generally agreed that "IAEA safeguards reflect an international consensus on what constitutes a high level of safety protecting people and the environment from harmful effects of ionizing radiation."

to adopt legislation for the effective protection of individuals, society, and the environment against radiological hazards.[60] It is in the setting of the conventions' periodic peer review meetings that this "incorporation by reference" bestows upon Agency safety standards a dynamic and normative quality (Handl, 2003). In no small measure, this is due to the adoption of revised guidelines regarding national reports under the two nuclear safety conventions.[61] In consequence, compliance with international standards is amenable to robust scrutiny as states are expected to provide a detailed article-by-article account in their national reports of how they implement and abide by relevant safety requirements, including those that govern EPR.

On the other hand, the model of "compliance control" applicable in the context of the Early Notification and Assistance Conventions, the "EPR Conventions" proper, is clearly less effective. To wit, it was not until 2014, at the Seventh (biannual) Meeting of the Representatives of Competent Authorities (CA) under the Early Notification and Assistance Conventions that national EPR reports were being presented for the first time. These biannual meetings provide more of a public forum for the discussion of EPR generic or topical issues rather than a mechanism for ascertaining individual states' compliance with specific EPR-related obligations under the conventions.[62] On top of it, some states parties oppose the very idea of national EPR reports on the ground that there are no reporting requirements under the Early Notification or the Assistance Conventions. Other states might object to such reporting as duplicative given existing reporting requirements under the safety conventions.[63]

Nevertheless, at the end of the day, these obvious differences in the robustness of the review process associated with the conventions directly bearing on EPR should not appreciably affect the ultimate normative effectiveness of the Agency's EPR safety standards. After all, EPR-related obligations under the two safety conventions and obligations under the two EPR Conventions proper overlap. The stringency of the review mechanism of the former therefore can compensate for the relative weakness of the latter, thereby ensuring that the Agency's EPR standards enjoy de facto normative status. This conclusion stands notwithstanding some criticism that IAEA EPR requirements are too numerous and detailed, thus rendering strict compliance with them virtually impossible (ENCO Final Report, 2013). For whether or not they serve as mere "benchmarks" for the identification of broad areas of good practice and opportunities for improvements as alleged (ENCO Final Report, 2013) there is no denying that they do create a normative pull which states will not be able to easily disregard or escape from.

---

[60] CNS, Preamble, Paragraph viii; and Joint Convention, Articles 4 and 11, respectively. In other words, they either inform the interpretation of the installation state's conventional obligations because they must be deemed expressly incorporated as such or, alternatively, because they generally reflect the degree of due diligence the installation state will have to apply in a particular situation.

[61] As to their latest versions, see IAEA (2015); and IAEA (2014).

[62] For the latest published summary account, see IAEA (2016).

[63] Seventh Meeting of the Representatives, *supra* note 49, at 7.

2.2 Other international, regional, and industry-inspired nuclear emergency preparedness and response efforts

Apart from this global setting, nuclear off-site EPR is being addressed also in various other international, regional, and industry-wide fora. These efforts not only reinforce and/or complement the global framework—often by adapting global standards and policies to specific local/regional conditions—but can also result in parallel structures and processes. For example, the OECD Nuclear Energy Agency[64] whose remit specifically identifies assistance with radiological emergency preparedness and management [65] organizes emergency exercises for member and nonmember states as well as international organizations.[66] While the NEA ostensibly seeks to minimize any overlap with, and duplication of, IAEA activities (Kovan (2005) at 32) its emergency exercises, such as its most recent INEX-5 exercise,[67] resemble very closely IAEA's own ConvEx exercises.

Given the long history of European cooperation on the peaceful use of nuclear energy,[68] it should come as no surprise that EPR has been a topic of considerable regulatory attention at the European level. Following the accident in Fukushima, the EU revised its nuclear safety directive.[69] However, while on-site EPR is being addressed in the directive, off-site EPR is not.[70] Rather, it is Council Directive 2013/59/EURATOM laying down basic safety standards for protection against ionizing radiation ("BSS Directive")[71] that determines key aspects of off-site EPR

---

[64] NEA's current membership consists of 31 countries in Europe, North America, and the Asia–Pacific region.

[65] See OECD/NEA, The Strategic of the Nuclear Energy Agency 2011–16, 21.

[66] Kovan (2005): "The International Nuclear Emergency Exercises (INEX) program was one of the NEA's responses to the Chernobyl accident. ... The first exercises dealt with the urgent early phase of an accident, within days of the occurrence of the release, concerned primarily with protecting people through such things as giving iodine, providing shelter, and evacuation."

[67] INEX-5 exercises were conducted in 2015–16. Its findings are summarized in OECD/NEA (2018).

[68] This cooperation began, of course, in 1957 with the adoption of the Euratom Treaty. For its latest version, see the Consolidated version of the Treaty establishing the European Atomic Energy Community, Doc. 2012/C 327/01, O.J. C 327/1, 26.20. 2012.

[69] Safety Directive, *supra* note 8.

[70] Nor was the latter a target of the European stress test. See European Nuclear Safety Regulators Group (2012). However, the European Nuclear Safety Regulators' Group (ENSREG) conducting the review did identify a strong demand for a European initiative on off-site emergency preparedness and recognized its importance in the follow-up to the Fukushima disaster. See ibid. at 49.

[71] Council Directive 2013/59/EURATOM of December 5, 2013, laying down basic safety standards for protection against the dangers arising from exposure to ionizing radiation and repealing Directives 89/618/Euratom, 90/641/Euratom, 96/29/Euratom, 97/43/Euratom, and 2003/122/Euratom.

throughout EU member states.[72] Implicit though the harmonization of off-site nuclear EPR in Europe thus might be, such a result is far from guaranteed.[73] Regional cooperation on radiological and nuclear EPR has also accelerated appreciably in Asia[74] within, in particular, the ASEAN Network of Regulatory Bodies on Atomic Energy (ASEAN, 2016) and among individual ASEAN member states.[75] Specifically, ASEANTOM, with the support from the EU and the IAEA (via its Technical Cooperation program) (IAEA, 2019; Commission Implementing, 2019), has developed a strategy for regional cooperation on radiological and nuclear EPR and an Action Plan for its implementation. "The strategy aims at bringing EP&R arrangements in ASEAN broadly in accord with best international practice within the next five to 10 years."[76] Similar efforts at regional harmonization of EPR have been underway within the ambit of the *Foro Iberiamericano* (Armonización), the Forum of Nuclear Regulatory Bodies in Africa,[77] the Arab Atomic Energy Agency (Mahjoub, 2015) as well as the Cooperation Council for the Arab States of the Gulf (IAEA, 2016). Building on a long history of cooperative ventures, the Nordic countries[78] too closely cooperate and coordinate national plans and responses, thereby forging a "Nordic approach" to nuclear EPR.[79] Finally, mention ought to be made here of industry initiatives, such as the program launched by the World Association of Nuclear Operators (WANO).[80] In 2011, WANO following a recommendation of its "Post-Fukushima Commission" decided to expand the scope of its peer reviews and other programs so as to "focus not only on preventing a nuclear event but also on mitigating the consequences of one if it should occur (WANO, 2011)." As a result,

---

[72] These include basic protective measures, emergency information, environmental monitoring, emergency management systems, response plans, and international cooperation. See preambular Articles 41–49, 69–71, and 97–98 and Annexes XI and XII.

[73] Indeed, as the Council of the European Union readily acknowledges, such an outcome will depend on member states' consistent transposition and implementation of the BSS Directive. See Council of the European Union(2015). See further OECD/NEA, Experience from the Fifth International Nuclear Emergency Exercise (INEX-5), *supra* note 67, at 45–46.

[74] As regards the early stage of these efforts, see, e.g., ASEANTOM (2013); ANSN (2013); Economic Research Institute for ASEAN and East Asia (2015).

[75] For further details on cooperative nuclear EPR efforts within the ASEAN region, see Yutthana (2018).

[76] See IAEA (2019) and Commission Implementing (2019). One key objective of the EU's support of ASEAN EPR efforts is to "establish a regional early warning radiation monitoring system capable of providing a high level of assurance that any significant increase in the level … of radiation … in ASEAN would be detected promptly." Commission Implementing (2019) at 10.

[77] See Nuclear Safety Review 2014, *supra* note 5, at Paragraph 124.

[78] These include Scandinavia, Finland, and Iceland. For details, see The Nordic Manual (NORMAN): Co-operation between the Nordic Authorities in Response to and Preparedness for Nuclear and Radiological Emergencies and Incidents, Revised August 2015.

[79] See Holo (2016). An example of this is the Nordic recommendations on operational intervention levels in a nuclear emergency. See Radiological Emergencies – Recommendations (2001); and Nordic Guidelines and Recommendations (2014).

[80] WANO was launched in the wake of the Chernobyl accident for the purpose of maximizing the safety and reliability of commercial nuclear powers plants worldwide.

emergency preparedness and severe accident management are now part of WANO's core mission. Other transnational actors, such as the International Federation of Red Cross and Red Crescent Societies (International Federation, 2015) may also play significant supportive roles in the management of nuclear off-site EPR.

Given these diverse efforts, the question might well be asked as to the degree to which there is a place for idiosyncratic regional EPR arrangements or standards. A 2013 report prepared for the European Commission (ENCO Final Report, 2013, at viii) seeks to address this matter in a European context. While acknowledging the utility of existing mechanisms for coordination between the European Commission and the IAEA to "ensure complementarity of activities, and respecting the principles of subsidiarity and proportionality (ENCO Final Report, 2013, at v), the Report nevertheless endorses a Europe-specific approach to at least some aspects of off-site EPR in preference to "reliance on work done at the international level by the IAEA."[81] Action at the European level, so it asserts, is "essential to ensure a consistent approach to compliance with EU [-specific] legislative requirements and a framework that is optimised for European ... social and economic conditions (ENCO Final Report, 2013)."

Arguably, the greater the socioeconomic and legal integration of a region, the stronger might be the argument in favor of a regional approach to nuclear EPR, especially if nuclear safety matters are already subject to region-specific standards. However, there are obvious technical and resource limits to how far any such regionalized approach might go.[82] Indeed, regionalization can be at odds with the very objective of harmonizing nuclear EPR measures internationally, a goal whose realization is essential for ensuring the credibility of the global nuclear safety regime.[83] When seen from this perspective, there is no denying that regional EPR standards will ultimately have to dovetail with global standards: "Broad compliance with ...

---

[81] ENCO Final Report (2013), at viii. The recommended measures include, inter alia, the expansion of EU-wide peer review to cover EPR; greater harmonization across Europe in respect of emergency planning zones and the introduction/removal of protective measures; as well as the development of a guidance document or codes of best practice regarding critical off-site EPR issues, such as cross-border arrangements. ENCO Final Report (2013). at vi.

[82] Thus ENSREG admitted that EU member states' compliance with the European Safety Directive's requirement for international peer review—Article 8e (1) of the Directive—was best achieved through cooperation with IAEA's IRRS program. Report of the European Nuclear Safety Regulators Group 14 (2015). "ENSREG also agreed that self-assessments should be based on IAEA IRRS practices, noting that IRRS missions look beyond the scope of the CNS and the Joint Convention obligations and Full Scope IRRS missions are beyond the scope of the Directive." See also Report of the European Nuclear Safety Regulators Group ENSREG 2017, Doc. HLG_p(2017-35)_155 ENSREG_REPORT_2017, 13; and Memorandum of Understanding between ENSREG and the IAEA for International Peer Review Missions to the EU Member States, June 28, 2011.

[83] For further discussion, see *infra* text at notes 126–142.

international safety standards in EPR [is] ... a key step to achieving harmonization."[84] In this sense, IAEA-promulgated EPR-related standards provide foundational normative guidance.[85]

---

3. Specific emergency preparedness and response policy challenges and international regulatory responses

3.1 Cross-border coordination of emergency preparedness and response: shared understandings and mutual trust

The overriding importance of consistency of off-site EPR policies and measures as between neighboring countries and across different regions was dramatically demonstrated first by the Chernobyl accident when response measures in affected regions of Western Europe inexplicably varied from country to country, at times even within the same country.[86] The issue surfaced with a vengeance also in the wake of the Fukushima accident: While the US government famously recommended that American citizens in Japan evacuate from an area of up to 50 miles from the stricken plant (The Washington Post, 2011), the Japanese authorities limited evacuation to an area with an initial radius of only 3, then 10, and eventually 20 km.[87] Needless to say, differences in national emergency plans for, perceptions of, and responses to a given major nuclear accident can be highly problematic because inevitably they raise awkward questions about their justifiability. In other words, the lack of a common understanding or approach to the management of off-site effects, certainly as between directly affected states—the accident state and neighboring risk-exposed state(s)—but also, as Fukushima clearly proves, as between accident

---

[84] International Conference on Global Emergency Preparedness and Response, October 19–23, 2015, Vienna Austria, Annex 2: President's Summary, 43, at 45. They provide more than the basic international framework for EPR, but "a solid basis for achieving ... harmonization." IAEA, Nuclear Safety Review 2015, Doc. GC(59)/INF/4, Paragraph 197.

[85] For an acknowledgment see Memorandum of Understanding ENSREG-IAEA, *supra* note 82, Article 4(1). The paramount importance of national EPR measures' consistency with Agency safety standards to avoid serious disruptions at the international level is broadly endorsed. See, e.g., Measures to strengthen international cooperation in nuclear, radiation, transport and waste safety, Resolution adopted on September 22, 2011 during the Seventh Plenary Meeting, IAEA Doc. GC(55)RES/9, Paragraph 83 (2011). See also Measures to strengthen international cooperation in nuclear, radiation, transport and waste safety, Resolution adopted on September 17, 2015 during the eighth Plenary Meeting, IAEA Doc. GC(59)/RES/9 (2015), preamble, Paragraph (bb), and Nuclear Safety Review 2015, *supra* note 84, at Paragraph 197.

[86] See, e.g., Handl (1988). Consider for example the maximum contamination values set for iodine-131 in milk. The United Kingdom and Sweden adopted a value of 2000 bequerels, whereas in Poland the limit was 1000, and in Hungary 500, in Austria 370, and in the state of Hesse, Germany, a mere 20. As to the possibility of different protective standards within one and the same country, see, e.g., McMahon (2011).

[87] See The National Diet of JAPAN, *supra* note 4, at 38.

state and other distant states (HERCA, 2013)—tends to undermine the very credibility of nuclear EPR generally.[88] A joint 2014 position paper of the Heads of European Radiological Protection Competent Authorities (HERCA)[89] and the Western European Nuclear Regulators' Association (WENRA)[90] warns that "differences can potentially have a significant effect, especially if the location of the emergency is close to an international border. Internationally, populations would feel unequally protected, depending on where they live (HERCA-WENRA, 2014)." The very same point is being made also in the ENCO Report that emphasizes that national differences in implementing EPR principles and objectives "are a source of misunderstanding, particularly among the public and politicians."[91]

Both the HERCA-WENRA Approach and the ENCO Report therefore recommend greater harmonization or better cross-border coordination of national protective actions,[92] a position that has long been advocated by the IAEA itself.[93] Of course, such harmonization or coordination is firmly rooted in EPR-related international legal obligations arising under the nuclear safety conventions. Thus, as noted, Article 16, paragraph 2 of the CNS requires that a nuclear installation state provide "appropriate information" to other states "in the vicinity of the nuclear installation" if it is likely to be affected by a radiological emergency. A risk-exposed Contracting Party that does not have a nuclear installation on its own territory, on the other hand, is obliged to "take the appropriate steps for the preparation and testing of emergency plans" for the protection of its own population and environment.[94] In the final analysis, these obligations clearly imply appropriately harmonized EPR measures as between accident state and risk-exposed neighboring state(s), as Contracting Parties to the CNS seem to readily acknowledge.[95] Indeed, the IAEA's Safety Requirements deliver a very similar

---

[88] See also French and Agryris (2014); and ENCO Report, *supra* note 64, at xiii.

[89] "Since its creation, HERCA identified the need for a harmonised approach on Emergency Preparedness and Response (EP&R) in Europe as a top priority." HERCA.

[90] WENRA is a nongovernmental organization comprised of the heads and senior staff members of all the national nuclear regulatory authorities of European countries with nuclear power plants.

[91] ENCO Report, *supra* note 64, at xii.

[92] See HERCA-WENRA Approach (HERCA-WENRA, 2014), at 15; and ENCO Report, *supra* note 64, at xii-xiii. See also Study on good practices (2019).

[93] See, e.g., IAEA (2002) and (2011, para. 82). The Nuclear Safety Review 2015, *supra* note 84, at Paragraph 197, reiterates that both practical insights gained from emergency exercises and discussion in EPR expert group meetings continue to confirm the importance of harmonized EPR arrangements worldwide.

[94] CNS, Article 16, Paragraph 3. Articles 6 (iv) and 13 (iv) of the Joint Convention contain similar provisions.

[95] For example, at Sixth Review Meeting of the Parties to the CNS contracting parties "noted the advantage of harmonizing the approach to severe accident analysis and the resulting emergency preparedness and response measures ... [as well as] the importance of harmonizing protective measures and trade measures to be taken during an emergency." Similarly, some contracting parties urged a complete and transparent exchange of information concerning possible transboundary effects of accidents "as this would facilitate the development of appropriate harmonized emergency preparedness and response measures." See 6th Review Meeting (2014).

message in that they call, in mandatory language, for "arrangements for coordination" among states to ensure consistency in assessing the accident situation and its radiological implications, and in the taking of protective or other response actions.[96] More specifically, they mandate "appropriate coordination across borders" to enable neighboring states with areas in threat category V[97] to develop "their own preparedness to respond to a transboundary emergency."[98] Encouragingly, IAEA member states are indeed, "showing increased interest in harmonizing their EPR arrangements based on the requirements of GSR Part 7 (IAEA, 2019)."

The ultimate objective of such transboundary coordination/harmonization cannot be complete cross-border uniformity. After all, emergency planning in general has traditionally been a matter of exclusive domestic jurisdiction, and while today's international (and regional) normative principles and standards do circumscribe states' discretion in devising domestic nuclear EPR policies and measures, they do not eliminate it completely. In short, national differences reflecting local political and economic priorities, societal sensitivity to risk, quite apart from regulatory authorities' idiosyncratic understanding and implementation of the EPR normative framework,[99] will persist. Thus, a more realistic goal might be for neighboring states to aim for what has been referred to as cross-border "shared technical understandings, coordination and mutual trust."[100] For such a transboundary relationship to materialize, the installation state as well as the neighboring state(s) must, first of all, faithfully and transparently comply with their various obligations arising under general international law,[101] the nuclear safety conventions, regional regimes, and IAEA safety standards in respect of the assessment of, and the sharing of information and mutual consultations bearing on, the transboundary nuclear risk involved. More specifically, a transboundary concertation of EPR would have to cover all in-advance determined critical parameters of off-site accident preparedness

---

[96] See, e.g., GSR Part 7 (IAEA, 2015), at Paragraphs 5.39 (requiring transboundary coordination where the emergency planning zone or distance extends across the border), and 6.13–6.14 (requiring that "governments ensure that arrangements are in place for the coordination of preparedness and response … at the international level," as appropriate).

[97] "Threat category V area" is an "area within the food restrictions planning radius, [i.e.,] the distance that could be affected by emergencies at a threat category I or II facility resulting in levels of ground deposition necessitating food restrictions consistent with international standards." IAEA (2003).

[98] GSR Part 7 (IAEA, 2015), at Paragraph 6.15.

[99] See, e.g., the ENCO Report (ENCO Final Report, 2013), at ix, which addressing the situation in the EU notes that "Member States often take different approaches to the practical implementation of essentially the same principles and objectives for off-site EP&R."

[100] HERCA (2015). See also Conference Report (2015), which refers to cooperation among the NORDIC countries and at the level of the European Union as "two examples of good regional cooperation in EPR" that promote trust and increase mutual understanding.

[101] For a discussion of such procedural obligations in the context of nuclear power activities, see, e.g., Handl (1992).

and response, including reference levels,[102] "observables" that trigger specific protective action[103] as well as criteria for the adjustment of emergency planning zones (EPZs) in response to evolving accident scenarios.[104]

The result of this process ought to be a cross-border alignment of national response measures that, importantly, should allow risk-exposed countries' authorities to follow with confidence—at least during the critical early phase of the accident—the accident country's lead with regard to protective measures.[105] It is for this reason that, for example, the EU Council has called upon member states to "include the alignment of protective measures along borders as a factor in emergency decision-making in accordance with … [an] optimised protection strategy (Council Conclusions, 2015)." By the same token, when EPR policies and measures of the installation state differ appreciably from those of other neighboring states notwithstanding transboundary information exchanges and consultations, the former must be prepared to explain publicly the basis for such divergence.[106]

3.2 Emergency planning zones

One of the principal tools for off-site EPR purposes is the establishment of EPZs and distances around nuclear installations.[107] States are expected to designate two planning zones for "category I"[108] nuclear facilities: A "precautionary action zone (PAZ)," an area in the immediate vicinity of the facility in which precautionary protective measures, such as the evacuation, sheltering in place and the administration of stable iodine to the population,[109] would be taken before or immediately after an off-site release of radioactivity to minimize exposure to the radioactive plume[110];

---

[102] "For an emergency exposure situation …, the level of dose, risk or activity concentration above which it is not appropriate to plan to allow exposures to occur and below which optimization of protection and safety would continue to be implemented." IAEA (2014).

[103] Such as "operational intervention levels." OILs are defined as the values of environmental measurements, in particular dose rate measurements, which set the threshold for the initiation of the different parts of the emergency plan and the taking of protective measures. *See* GSR Part 7 (IAEA, 2015), at Paragraph 4.28.

[104] That is, in response to changing plant conditions and environmental monitoring results.

[105] See the HERCA-WENRA Approach, *supra* note 83, at 16.

[106] See Requirement 22 of GSR Part 7 (IAEA, 2015), at Paragraph 6.14.

[107] "[P]lanning areas' nature and size are an important basis for the implementation of protective measures and the development of strategies." Planning areas for emergency response near nuclear power plants: Recommendation by the German Commission on Radiological Protection 8 (2014).

[108] "Facilities, such as nuclear power plants, for which on-site events (including very low probability events) are postulated that could give rise to severe deterministic health effects off the site, or for which such events have occurred in similar facilities." IAEA (2007).

[109] See IAEA (2007), Annex V, 95−103.

[110] The principal goal of measures in the PAZ is to prevent severe deterministic effects, i.e., effects that can be related directly to the radiation dose received. The severity increases as the dose increases. A deterministic effect typically has a threshold below which the effect will not occur. The effect is deemed severe "if it is fatal or life-threatening or results in a permanent injury that reduces the quality of life." See IAEA (2013).

and an adjoining "urgent protective action planning zone" (UPZ), in which, following an environmental release, measures would be taken to minimize radiation exposure along the expected ingestion pathway.[111] Additionally, the IAEA Safety Requirements call for the establishment of an extended planning distance (EPD) beyond the UPZ to permit timely intervention to reduce the risk of stochastic health effects, and, beyond that, an ingestion and commodities planning distance (ICPD) over which response action for the protection of the food chain, water supplies, and commodities other than food (and the public potentially affected by such commodities) could be taken following a significant release of radioactivity.[112] Both EPD and ICPD "are to be established at the preparedness stage for the identification of areas in which actions may need to be taken during the response but for which only limited arrangements are put in place in advance."[113]

Clearly, the nature and size of EPZs and distances are a matter of international concern. This is self-evidently so whenever the planning zones or distances extend across an international border.[114] This may be the case also—albeit less obviously so—in situations in which emergency planning distances do not reach and let alone cross-international boundaries. For, as pointed out before, different national configurations of EPZs/emergency planning distances irrespective of their proximity to an international border tend to be understood as signaling different protection levels and hence are likely to raise doubts about nuclear EPR in general.[115]

Prior to the accident, a 10-km EPZ was in place around the Fukushima Daiichi nuclear power plant. However, as the IAEA Fukushima Report concludes, "the extent of [the zone] ... did not take into account the potential for a severe accident. In addition, provisions were not in place to extend relevant protective actions beyond

---

[111] GSR Part 7 (IAEA, 2015), at 29–31. Measures to be taken in the UPZ after a release of radioactivity aim at reducing the risk of stochastic effects, i.e., effects that occur on a random basis independent of the size of dose. While the effect has no threshold, the chances of seeing the effect increase with dose.

[112] Ibid. at Paragraphs 5.38 (iii)–(iv).

[113] IAEA, Actions to Protect the Public, *supra* note 85, at 20.

[114] Luxemburg, Germany, Croatia, Romania, Hong Kong, and Canada are among several countries with territory within the EPZ of neighboring countries.

[115] The ENCO Report, while warning against an oversimplistic equation of size of EPZ with level of protection afforded, nevertheless admits that such conclusions are inevitable and therefore will be a source of public and political concern." *See* ENCO Report, *supra* note 64, at 25. See also Planning areas for emergency response near nuclear power plants, *supra* note 107, at 15.

the emergency planning zone."[116] In short, in respect of both the size of the EPZs and corresponding specific protective measures,[117] Japan's nuclear EPR proved to be deficient.[118] Shortly after the accident, a number of countries began to reevaluate the adequacy of their EPZs.[119] At the same time, an IAEA group of experts concluded that "[a]n internationally agreed calculation methodology ... [was] needed for determining the optimal size of emergency planning zones. Emergency planning zones ... [were] to be redefined to take into account the experience from the accident at the Fukushima Daiichi nuclear power plant."[120] As regards the EPR situation in Europe, the authors of the HERCA-WENRA Approach suggest that "an accident comparable to Fukushima would require protective actions such as evacuation up to 20 km and sheltering up to 100 km. These actions would be combined with the intake of stable iodine."[121]

---

[116] IAEA, Technical Volume 3/5, *supra* note 14, at 96.

[117] Under deteriorating conditions at the stricken Fukushima Daiichi plant, Japanese authorities were repeatedly forced to expand the initial evacuation area out to 20 km. Eventually, protective measures were ordered also for residents in an area between 20 and 30 km from the plant: Although at first only ordered to shelter-in-place, these residents were eventually urged to voluntarily evacuate. See The National Diet of JAPAN, *supra* note 4, at 38. However, the accident produced also radioactive hotspots further afield. For example, at Iitate, 40 km from the plant, levels of caesium-137 were as high as 18 MBq/m^2, well above the level at which evacuation would be deemed advisable. See New Scientist (2011).

[118] See The Fukushima Daiichi Accident, *supra* note 4, at 84−90.

[119] See generally NEA, the Fukushima Daiichi Nuclear Power Plant Accident, *supra* note 4, at 28. See further Office for Nuclear Regulation (2011). Clearly, the advent of new nuclear power systems with advanced safety features, such as the Gen-III European power reactor (EPR) or the AP 1000, raises the issue of whether present EPZs might have to be revised. Some experts have suggested that these new reactors would permit a drastic reduction of EPZs. For example, B Tomić, *supra* note 18, mentions a maximum 800-m planning zone for long-term protective measures. However, countries that presently host or will host such reactors, such as France, Finland, or China, are not known to have made any adjustments of their traditional, Gen-II related EPR regulations or indicated a willingness to do so.

[120] Main Conclusions from the Workshop on Sharing Lessons identified from Past Responses and Exercises, April 23−27, 2012, Vienna, in IAEA Report on Preparedness for a Nuclear or Radiological Emergency in the Light of the Accident at the Fukushima Daiichi Nuclear Power Plant, Annex A, 40 (2013).

[121] HERCA-WENRA Approach (HERCA-WENRA, 2014), at 9. More specifically, the study recommends urgent protective actions as well as a minimum common level of preparation for action, namely evacuation up to 5 km around nuclear power plants, and sheltering and iodine thyroid blocking (ITB) up to 20 km; and a general strategy to extend evacuation up to 20 km, and sheltering and ITB up to 100 km.

International nuclear law does not prescribe a specific size for the planning areas concerned. While the IAEA suggests a mere 3−5 km radius for the PAZ and a 5−30 km radius for the UPZ,[122] many nuclear installation states, including the United States,[123] have opted for somewhat larger PAZs covering areas of a 10−16 km radius.[124] The IAEA recommended maximum distances for the EPD and ICPD of large nuclear power plants, are 100 and 300 km, respectively.[125] As regards size, emergency planning areas are likely to vary from country to country. After all, emergency zones/distances will need to be based on site-specific factors, such as topography, population density, infrastructure, and so on. Moreover, as the ENCO Report notes, despite common principles underlying their establishment in most countries in Europe, sizes of EPZs in practice differ considerably—reflecting different value judgments about what is reasonable to plan for in a detailed manner.[126] Nevertheless, comparability of national approaches—based on compliance with IAEA's fundamental organizational principles[127]—would be desirable because it would bolster public confidence in EPR programs. This is true even though, admittedly, emergency planning areas' size per se is not necessarily a reliable indicator of the level of protection that could be expected within the areas concerned.[128]

---

[122] See IAEA, Actions to Protect the Public, *supra* note 110, at 22.

[123] See 10CFR §50.47 (c) 2: "Generally, the plume exposure pathway EPZ for nuclear power plants shall consist of an area about 10 miles (16 km) in radius and the ingestion pathway EPZ shall consist of an area about 50 miles (80 km) in radius." Post-Fukushima, the US Nuclear Regulatory Commission opposed an increase in the size of the present plume exposure pathway zone from 10 to 25 miles, establishing a new zone but with less stringent requirements from 25 to 50 miles around reactors, expanding the existing ingestion pathway zone from 50 to 100 miles. See Petition for Rulemaking to Improve Emergency Planning Regulations (10CFR 50.47), at http://www.nirs.org/reactorwatch/ emergency/petitionforrulemaking22012.pdf, visited September 16, 2016. The Commission maintained that "the current size of the emergency planning zones … [was] appropriate for existing reactors and that emergency plans … [would] provide an adequate level of protection of the public health and safety in the event of an accident at a nuclear power plant." See https://www.federalregister.gov/ articles/2014/04/09/2014-07981/emergency-planning-zones, visited September 16, 2016.

[124] See, e.g., G. Handl, *supra* note 101, at 30−35; and Kubanyi et al. (2008).

[125] See IAEA, Actions to Protect the Public, *supra* note 110, at 22. These calculations are for nuclear power plants with a capacity of more than 1 GW (th).

[126] ENCO, *supra* note 64, at 25.

[127] Such as its dose-related approach based on representative source terms. See Planning areas for emergency response near nuclear power plants, *supra* note 107, at 15, and see generally Criteria for Use, No. GSG-2, *supra* note 57.

[128] Indeed, simplistic assumptions about the relevance of EPZs have rightly been labeled as falling "for the proximity trap." See The Guardian (2019).

Three final observations might be offered here. First, the establishment of national emergency planning areas must cover accident scenarios that are highly improbable, rather than merely possible. One of the most significant planning mistakes affecting the way in which the Fukushima accident evolved was the fact that "a reactor-core damaging event at a nuclear power plant in Japan was considered implausible" and hence was not taken seriously (National Research Council, 2014). Prompted by the implications of this shortcoming in Japan's emergency preparedness, some countries have since begun to shift the focus of their EPR policies to "reflect more closely an accident's potential impact rather than its likelihood (OECD/NEA, 2016)." Second, in line with this reorientation of EPR installation states ought to employ the tool of "emergency planning distances" as specified in the IAEA safety requirements. While the EPD or ICPD distance parameters are merely recommended, the use of the concepts as such for EPR purposes is mandatory. Unfortunately, today many states' EPR policies and measures have yet to come into compliance with this requirement.[129] Finally, the advent of new nuclear reactor systems with advanced safety features,[130] such as the Gen-III + European power reactor (EPR), the VVER-1200, the AP 1000, and so on,[131] raises the issue of whether present EPZs ought not to be revised. Some experts have indeed suggested that these enhanced reactors would permit a drastic reduction of EPZs.[132] However, countries that already deploy or will soon host such reactors, including Russia, China, France, the United Kingdom, and Finland, are not known to have adjusted their national EPZ regulations/policies or indicated a willingness to do so.[133]

---

[129] See, e.g., ENCO Report, *supra* note 64, at 88 (noting the absence in most European countries of any planning zones for purposes of food restrictions).

[130] Whereas so-called generation IV reactors are still—mostly—only on the drawing board, several generation III and III + reactors are online now. See generally Reinberger et al. (2019).

[131] For an overview of generation III + reactors, see also Advanced Nuclear Power Reactors (2020).

[132] For example, B. Tomić, *supra* note 18, refers to a maximum EPZ of 800 m for long-term protective measures.

[133] According to Damien Didier, Head of Atmospheric Transfers Modeling Section (BMCA) of France's Institut de Radioprotection et de Sûreté Nucléaire, EPZs for these advanced reactor systems are similar to those for generation II reactors, despite the safety improvements associated with the former. Message of November 17, 2018, to Philip Andrews-Speed, Energy Studies Institute, National University of Singapore, *on file* with the author. However, the specific EPR implications of generation IV reactors have been under review for some time. See, e.g., Report of the Technical Meeting, *supra* note 10, 3. While acknowledging that EPR criteria of GSR Part 7 "are fully applicable to the definition of EPR arrangements needed for next generation reactors," the Report also notes the need for new next-generation reactors—related guidance on "criteria/factors to be considered … [in] the decision about the size of emergency planning zones and distances …." Ibid. at 8—9. See also U.S. Nuclear Regulatory Commission, "Emergency Preparedness Requirements for Small Modular Reactors and Other New Technologies," 82 Federal Register 52862, November 15, 2017 (proposing a graded approach in "right-sizing" EPZs for small modular nuclear reactors).

3.3 Event reporting and information sharing

An essential precondition for the successful management of potential or actual off-site radiological effects in a transboundary context is the cross-border flow of timely, reliable, and accurate information regarding the nature and scope of the accident. [134] As already noted, the installation state is under an international conventional and customary legal obligation to deliver such information to the risk-exposed neighboring state(s) and may well be so obliged also pursuant to regional (as in the case of EU member states)[135] or bilateral legal frameworks bearing on EPR. A corresponding conventional obligation arises also toward the IAEA and the regional authorities concerned. Clearly, several factors determine the ultimate effectiveness of this transboundary communications process,[136] including the operational integration of regional[137] and global[138] emergency information exchange platforms.[139] It is, however, the degree to which the specific factual circumstances triggering the obligation of transboundary emergency notification are defined exactly and objectively that stands out as being of critical importance. Unsurprisingly, the issue has been the object of much international regulatory attention and hence will also be the focus of inquiry in the present context.

Notoriously, Article 1, paragraph 1 of the Convention on Early Notification establishes multiple thresholds for the obligation to notify to arise by tying the obligation to a triple affirmative determination by the installation state regarding the consequences of the emergency: first, that a release of radioactive material occurs

---

[134] See generally Study on good practices, *supra* note 92.

[135] See Council Decision of 14 December 1987 on Community arrangements for the early exchange of information in the event of a radiological emergency (87/600/Euratom); and Agreement between the European Atomic Energy Community (Euratom) and nonmember states of the European Union on the participation of the latter in the Community arrangements for the early exchange of information in the event of radiological emergency (Ecurie), Official Journal C 102, 29/04/2003.

[136] Of course, the very quality (timeliness, comprehensiveness, and accuracy) of the data transmitted and the mode of transboundary communication. However, a detailed analysis of these various factors would far exceed the necessarily limited scope of this paper. For a discussion, see instead Operations Manual for incident and Emergency Communication, *supra* note 51.

[137] For a discussion of the European system, see *infra* text at notes 188–195.

[138] Namely, IAEA's Unified System for Information Exchange in Incidents and Emergencies (USIE). The IAEA website at https://iec.iaea.org/usie/actual/LandingPage.aspx, visited September 13, 2019, describes USIE as "an IAEA web portal for Contact Points of States Parties to the Convention on Early Notification of a Nuclear Accident and the Convention on Assistance in Case of a Nuclear Accident or Radiological Emergency and of IAEA Member States to exchange urgent information during nuclear and radiological incidents and emergencies, and for officially nominated INES National Officers to post information on events rated using the International Nuclear and Radiological Event Scale (INES)." See further HERCA (2014)

[139] See, e.g., Council of the European Union (2013), 42 (discussing the EURDEP's use on a global level by the IAEA).

or is likely to occur; second that it has resulted or may result in an international transboundary release; and last, that such a release would be of radiological safety significance for another state.[140] Especially during the first few hours of an incident at a nuclear facility, however, the installation state might simply not be in a position to make these determinations or, equally plausibly, be reluctant to do so for political reasons. Indeed, delays in transboundary or international notifications have been a rather more common phenomenon, the most notorious example of which is, of course, the USSR's failure to notify the international community for several days of the accident at the Chernobyl nuclear power plant,[141] which in turn prompted the launch of the Convention on Early Notification. Thus, reliance on the installation state authorities' subjective judgment is inherently problematic and potentially represents a serious obstacle to effective emergency communications.[142] Unsurprisingly, at the first post-Fukushima biannual meeting in 2012 of the representatives of competent authorities identified under the Early Notification Convention and the Assistance Convention (IAEA, 2012), two "nonpapers" by Japan and Russia called attention to this problem.[143] Specifically, the Japanese delegation urged the establishment of an international system—possibly through a resolution of the IAEA General Conference—which would have required the installation state to notify the IAEA in the event of a nuclear accident, even before the state had adjudged that the incident fell within the ambit of the notification-triggering Article 1.1 of the Convention.[144]

While the desirability of "objectivizing" the Convention's emergency notification threshold might be widely recognized, it has not—yet—led to a formal adjustment of the global legal regime. Responding to this matter, the Agency instead recommends prompt action of a kind by the installation state that is in line with some of the more advanced bilateral transboundary notification schemes. Thus whenever, because of its closeness to an international border, a facility's EPZ extends into a neighboring country's territory, an emergency notification would be expected to be sent to the neighboring state's authorities (as well as the Agency's IEC) at the same time as it would be sent to the installation state's own off-site authorities.[145] As regards facilities further inland, notifications are expected to be sent "forthwith (i.e., within less than 2 h) after the declaration" of a nuclear

---

[140] "This Convention shall apply in the event of any accident involving facilities or activities … from which and which has resulted or may result in an international transboundary release that could be of radiological safety significance for another State."

[141] For a recent detailed account, see Higginbotham (2019).

[142] See already Lang (1988); and N Pelzer, "Learning the Hard Way: Did the Lessons Taught by the Chernobyl Nuclear Accident Contribute to Improving Nuclear Law?" in OECD/NEA, International Nuclear Law, *supra* note 25, at 73, 80.

[143] See ibid. at 42 and 48, respectively.

[144] Ibid. at 45−46.

[145] Operations Manual For Incident and Emergency Communication, *supra* note 46, at 10.

emergency and follow-up information not later than 4 h after the initial transboundary notification.[146] These Agency clarifications obviously go some way to assuage concerns about the Convention's notification threshold, but they are in the nature of mere recommendations. In other words, the challenge of securing in legally binding terms the international emergency notification system's adequately prompt activation remains unanswered. However, many countries have striven for greater objectivity and comprehensiveness in specifying the threshold in bilateral agreements.[147] Ideally, neighboring countries ought to provide each other automatically—i.e., without the need of intercession by installation state authorities—with comprehensive real-time information on critical parameters reflecting the status of the nuclear installation concerned, as well as conditions on-site and off-site. Unfortunately, such sharing of information or complete openness to outside scrutiny by potentially affected neighboring states (or relevant regional bodies or the IAEA) is still far from being the rule today.[148]

At the European regional level, the notification threshold at law itself is relatively high and remains pegged to the decision by the installation state "to take measures of a widespread nature to protect the general public."[149] However, as the ENCO Report concludes, most European countries "have mechanisms in place to ensure timely notification of emergencies to neighboring countries over and above obligations under the Convention on Early Notification and the Community's Urgent Radiological Information Exchange System."[150] Many neighboring countries cooperate also extensively through bilateral commissions or groups of technical experts whose remit includes EPR.[151] Moreover, specific bilateral notification arrangements, such as they are, are complemented by the European Union Radiological Data

---

[146] Ibid. at 10—11.

[147] For details, see G. Handl, *supra* note 101, at 67—73; and Handl (2016).

[148] Even recent attempts to improve upon bilateral transboundary information exchanges, such as the 2015 agreement between the Norwegian Radiation Protection Authority and Rosatom, may fail to eliminate all subjective elements in the assessment of the circumstances that trigger the transboundary notification obligation. For details, see G. Handl, *supra* note 147, at 339—340.

[149] See Article 1 of Council Decision 87/8600/Euratom and Article 2 of the Agreement between the European Atomic Energy Community (Euratom), *supra* note 135.

[150] ENCO Report, *supra* note 64, at Appendices, 106. For example, emergency notification arrangements between Switzerland and Germany are said to provide decision-making bodies of the risk-exposed state with "almost the same access to information" as authorities of the installation state. Ibid. at 197. See Vereinbarung zwischen dem Schweizerischen Bundesrat und der Regierung der Bundesrepublik Deutschland über den radiologischen Notfallschutz, May 31, 1978, AS 1979 312; and Notenaustausch vom 25. Juli 1986 zwischen der Schweiz und der Bundesrepublik Deutschland betreffend die Durchführung der Vereinbarung vom 31. Mai 1978/15. Februar 1980/25. Juli 1986 über den radiologischen Notfallschutz, AS 1988 781.

[151] As regards, for example, Germany's cooperation with neighboring states, see, e.g., Bundesministerium für Umwelt, Naturschutz, Bau und Reaktorsicherheit, Bilaterale Zusammenarbeit auf dem Gebiet der kerntechnischen Sicherheit, at www.bmub.bund.de/P297/, visited March 12, 2016.

Exchange Platform (EURDEP), which makes national radiological monitoring data from most European countries available in near-real time.[152] It is, however, equally true that some of these transboundary arrangements, be they of a bilateral or European regional nature, lack a solid legal foundation. For example, participation in EURDEP is legally required of EU member states, whereas participation by non-EU countries is voluntary.[153] On top of it, some countries' monitoring systems might still be simply incompatible with common European data exchange formats.[154] Additionally, some elements of present-day bilateral transboundary notification and emergency information exchange arrangements rather than being binding in a formal legal sense represent informal cross-border understandings between regulatory agencies or simply individuals.[155] In the end, they may well prove less resilient than expected, especially during the critical but potentially confusing early hours of a nuclear accident. Clearly, a binding rules−based approach throughout the wider European region would be preferable. Only such a system or common legal framework, European countries seem to agree, would be capable of ensuring the "instantaneous exchange of information (notifications, alerts, forecasts, summary of measured data, plant parameters, countermeasures)" in a radiological incident[156]

At the global level, the IAEA's Incident and Emergency System is the lynchpin of the Convention on Early Notification−based emergency notification network. Since Fukushima, it has seen a remarkable expansion and transformation into a radiological emergency communication, evaluation, and response system. A critically important step in this evolution was the introduction in 2011 of the Unified System for Information Exchange in Incidents and Emergencies (USIE), a common single web platform for the exchange of information.[157] A second tool that complements USIE is the International Radiation Monitoring Information System (IRMIS),[158] which launched in

---

[152] EURDEP is currently used by 38 European countries—including all 28 EU member states as well as Norway, Switzerland, Belarus, Russia, Azerbaijan, Turkey, etc.—for the continuous exchange of data from their national radiological monitoring networks. During radiological emergencies, the rate of data delivery will be hourly. European Radiological Data Exchange Platform, at https://remon.jrc.ec.europa.eu/, visited February 3, 2020. Its main aim, as the EURDEP website explains, "is to notify and inform competent authorities and the general public during the early phase of a large-scale accident with release of radioactivity to the atmosphere as early and extensively as possible."

[153] See ibid.

[154] A case in point is Ukraine's present radiation monitoring system. For details, see Commission Implementing Decision of 12.7.2019 (WANO, 2011), at 11−12.

[155] See also ENCO Report, *supra* note 64, at Appendix 197.

[156] Ibid.

[157] USIE facilitates the exchange of notifications and information between competent national authorities. It also allows them to request information or international assistance.

[158] IAEA (2020): "In an emergency situation, through the aggregation and display of routine monitoring data, IRMIS can quickly be expanded into a powerful emergency communications tool that aids the assessment and prognosis conducted by the Accident State, the IAEA and supporting States."

2016 (Harvey) IRMIS does not represent an early warning system but rather serves as a decision support tool, enhancing situational awareness. It facilitates "the reporting and visualization of large quantities of radiation monitoring data" during an emergency and thereby enables "near-real-time monitoring of the evolving radiological situation worldwide (Harvey, at 3)." The Agency has also been in the forefront of developing a common international standard information exchange format—IRIX— to "facilitate the web-based exchange of relevant emergency information and data among national and international organizations that are involved in the response to nuclear and radiological emergencies (IAEA, 2020)." Significantly, IRIX eliminates the need to convert and transcribe information between different data presentations and formats that might otherwise impede the "compilation, exchange, and processing of emergency information and data (IAEA, 2020).

Finally, mention must be made here of a reverse flow of emergency information, this time from the IAEA to the installation (accident) state (as well as to other states). The Fukushima accident highlighted in dramatic fashion the need for a single institutional voice at the global level capable of characterizing accurately and authoritatively for the world at large the accident and its likely progression (IAEA, 2015). Part of the explanation for the confusing, often contradictory, multiple-sourced flow of information about the status of the Fukushima Daiichi reactors and the likely progress of the accident,[159] was Japan's difficulty in meeting and managing the demand for information by decision-makers and the public.[160] The 2011 Action Plan on Nuclear Safety Agency recognized some of these difficulties as symptomatic of the "fog" during the early stages of a nuclear accident.[161] It therefore called upon the IAEA to "provide Member States, international organizations, and the general public with timely, clear, factually correct, objective and easily understandable information during a nuclear emergency on its potential consequences, including analysis of available information and prognosis of possible scenarios based on evidence, scientific knowledge, and the capabilities of Member States."[162] A critical part of this mandate is, in the end, the Agency's delivery of accident-related information for dissemination among member states, the public and the media, messaging that is harmonized between the accident state, the IAEA Secretariat, and any member state

---

[159] Apart from the difficulty of acquiring real-time accurate data, Japanese authorities also experienced major problems in assessing the information that was available and in offering a global view of the accident, its off-site impact, and likely future course of development. For a discussion of these communication failures, see, e.g., National Research Council, Lessons Learned, *supra* note 4, at 221–222; and IAEA, Technical Volume 3/5, *supra* note 14, at 85–93.

[160] In other words, to be effective, emergency communications require authoritative interpretation of data, and the presentation of the informational output in a contextually "proper perspective." See IAEA Report on Assessment and Prognosis (IAEA, 2015), at 13.

[161] At the time of the accident, the Agency's functions did not include "providing a prognosis of the potential evolution of an accident or an assessment of the possible consequences." See IAEA, Technical Volume 3/5, *supra* note 14, at 158.

[162] IAEA Action Plan, *supra* note 15, at 6.

supporting the process (IAEA, 2019, 2020). Today, this Agency function is generally recognized as providing a potentially indispensable service to national decision-makers facing a nuclear emergency.[163]

3.4 Validation of national emergency preparedness and response through international peer review and emergency exercises

To ensure the ultimate effectiveness of nuclear off-site emergency arrangements, EPR planners must engage in a process of continuous improvement through the sharing of relevant information and experience as well as related consultations with other decision-makers and stakeholders, including the public. Indispensable elements in this process are periodic peer reviews and the testing of EPR arrangements in realistic emergency exercises, both at the local, national, and international levels.[164] Indeed, it is by now a commonplace that the importance of independent outside reviews and practical testing of EPR "cannot be overemphasized."[165] Periodic testing and peer reviews of EPR are key proposals of the IAEA Action Plan on Nuclear Safety.[166] Indeed, the nuclear safety conventions themselves establish a clear legal obligation for the installation state to ensure that off-site emergency plans "are routinely tested" and "cover the activities to be carried out in an emergency"[167]; or "are tested at an appropriate frequency."[168] The IAEA Safety Requirements, moreover, call not only for "training, drills and exercises" for EPR,[169] but they stipulate also, as part of a quality management program, that national EPR arrangements undergo "periodic and independent appraisals," including international appraisals, such the IAEA-organized EPREV.[170] Apart from local and national testing of EPR, there is also an obvious, often pressing need[171] for transboundary nuclear emergency exercises.[172] Many of these will be conducted

---

[163] For example, the INEX-5 workshop lauded that the Agency's assessment and prognosis tools were "helpful for decision makers as the statements generated through the use of these tools … [were] clear and are presented in an easy-to-be-understood language. This aid[ed] the technical experts in their role with the comprehension of the output by the decision makers. It was noted that this set of tools might be particularly useful for non-nuclear countries." See OECD/NEA, *supra* note 67, at 46.

[164] This is clearly one of the many important lessons of the Fukushima accident. For further details, see, e.g., Institute of Nuclear Power Operations, *supra* note 4, at, 30, 35.

[165] International Conference, *supra* note 84, at 14; and IAEA 2006, 1: "Emergency response exercises are a key component of a good emergency preparedness program. They can provide unique insight into the state of preparedness of emergency response organizations."

[166] *Supra* note 15, at 3.

[167] CNS, Article 16, Paragraph 1.

[168] Joint Convention, Article 25, Paragraph 1.

[169] GSR Part 7 (IAEA, 2015), Requirement 25, 56–57.

[170] Ibid. at 58.

[171] Note in this respect the critical findings by Nuclear Transparency Watch 2015, 54–55.

[172] Note in this context the EU Council's call for neighboring Member States to undertake "joint training sessions and nuclear emergency exercises representative of real emergency situations …" See *supra* note 75, at 8; and Study on good practices, *supra* note 92, at 10–11.

among neighboring countries bilaterally; others multilaterally on a regional[173] or global level. As already noted, these multilateral drills include tests organized by the European Commission,[174] the OECD/NEA (INEX exercises),[175] of course, the IAEA/IACRNE (ConvEx exercises)[176] as well as WANO.[177] They range from table-top exercises to field drills, can run for several hours or days, and are either stand-alone exercises or may represent the combined efforts of several sponsoring organizations.

The importance of periodic peer reviews—overwhelmingly endorsed by the regulatory community and the nuclear industry in respect of the safety of nuclear power plant operations in general[178]—applies with equal force to EPR planning for off-site effects.[180] Thus, the Action Plan on Nuclear Safety recognizes the need for strengthening IAEA peer reviews of EPR; for enhancing their transparency through the publication of summaries and, with the consent of the state concerned, the full results[180]; and for states to voluntarily and regularly submit to them as well as follow-up reviews.[181] Specifically, the Action Plan calls upon states "to conduct … regular reviews of their emergency preparedness and response arrangements and capabilities,

---

[173] As a case in point is the three-nation exercise simulating an accident at the Cattenom nuclear power plant, involving Luxemburg, France, and Germany during 2011−13. For details, see France, CNS, 6th National Report for the 2014 Review Meeting (June 2013), 132.

[174] For example, the October 2015 ECURIE exercise tested the European-wide emergency information system.

[175] See *supra* text at notes 70-74.

[176] IACRNE, for which the IAEA provides the secretariat functions, coordinates large-scale international emergency exercises with the participation of a host state, of other states and various international governmental organizations concerned with nuclear EPR. For details on ConvEx-1 through 3 exercises, see *supra* text at notes 51−54. For details on the latest and to date largest ConvEx-3 exercise postulating a loss of coolant and release of radioactive material accident at Hungary's Paks Nuclear Power Plant, see "Convention Exercise Level 3: IAEA Tests Global Emergency Response in Largest Nuclear Accident Simulation to Date," IAEA-Incident and Emergency Centre, EPR Insights September 3, 2017.

[177] In 2013, WANO organized a Scandinavian-Baltic region exercise. See WANO Moscow Centre Regional Crisis Centre in Action, http://www.wano.info/en-gb/mediaandevents/pressreleasesandannouncements/Pages/WANO-Moscow-Centre-Regional-Crisis-Centre-in-action.aspx, visited March 14, 2016.

[178] See Handl, *supra* note 24, at 203−209.

[179] On a global level, two types of international peer reviews are of special relevance in the present context, namely IAEA's dedicated emergency preparedness review (EPREV) and, at least to some extent, its Integrated Regulatory Review Service (IRRS). The latter provides reviews of the effectiveness of national regulatory infrastructure, including EPR organizations, against applicable international guidelines and best practices.

[180] Note also the statement of the US delegate at the IAEA Board of Governors Meeting, March 7−11, 2016, at http://vienna.usmission.gov/160307safety.html, visited October 27, 2016: "In order to continue to enhance our nuclear and radiation safety programs, as well as emergency preparedness and response readiness, we urge all Member States to request and participate in not only peer review missions, but also in the related follow-up missions, and to publish the results of those missions to promote transparency and openness."

[181] IAEA Action Plan, *supra* note 15, at 3.

with the IAEA Secretariat providing support and assistance through Emergency Preparedness Review (EPREV) missions, as requested."[182]

In contrast to the *periodic testing* of EPR, there exists no equivalent international conventional or other clear legal basis for mandatory international peer reviews of emergency plans, the exception being the European Nuclear Safety Directive that mandates an international peer review of national EPR arrangements, albeit only in the immediate aftermath of an accident with off-site consequences.[183] However, this does not necessarily imply that today states are entirely free to decide whether or not to seek or submit to independent outside appraisals of their emergency plans. For, as noted, a state's international legal obligation to seek periodically an independent international assessment of its EPR plans can be construed to flow from Requirement 26 of IAEA's revised EPR Safety Standards.[184] Admittedly, there has been some push-back against the idea of EPR assessments taking the form of a "prescriptive appraisal of [States'] emergency arrangements against a standard with no binding nature."[185] Also, the ENCO Report in recommending that, analogous to the mandatory peer reviews under the European Nuclear Safety Directive,[186] the European Commission propose legislation for periodic EU-wide peer reviews of national off-site EPR arrangements[187] seems to assume the absence of an international legal obligation to this effect. At the same time, it is worth noting that at the 2015 International Conference on Global Emergency Preparedness and Response, the chairman of the international expert meeting recommended that the contracting parties to the CNS use the Convention's peer review process "to ensure the continuous enhancement of EPR to a nuclear or radiological emergency."[188]

In sum, as regards the issue of whether today the periodic, routine peer review of national EPR corresponds to an international legal obligation, the picture that emerges is somewhat contradictory. Nevertheless, what can be said with some degree of certainty is that while peer reviews of national EPR within the context of the CNS review meetings are likely to prove unproblematic, the opposite may well be true of similar exercises in different international settings or forums. Moreover, for the time being at least, special in-country, on-site EPR appraisals, such as through IAEA's EPREV missions, are likely to remain grounded in states' decision to voluntary seek out to this service. The independent international peer review,

---

[182] Ibid.

[183] Article 8 (e), Paragraph 4: "In case of an accident leading to situations that would require off-site emergency measures or protective measures for the general public, the Member State concerned shall ensure that an international peer review is invited without undue delay."

[184] GSR Part 7(IAEA, 2015), at 57.

[185] Statement by France, Report of the Seventh Meeting of Representatives, *supra* note 49, at 10, Paragraph 53.

[186] See Article 8(e) of the Safety Directive, *supra* note 8. It also requires that the results of the review be reported to the member states and the Commission.

[187] ENCO Report (ENCO Final Report, 2013), at, xii.

[188] International Conference, *supra* note 84, at 44.

however, is a cornerstone of the global nuclear safety regime, in general, and states' emergency planning, in particular. Therefore, states' discretion in seeking or accepting to undergo such periodic reviews, to the extent it does exist today, ought to be severely curtailed, if not eliminated.

4. Nuclear emergency assistance: global, regional, and bilateral arrangements

One of the most important factors bearing on the mitigation off-site effects during the early stages of a nuclear accident may be the availability of transboundary emergency assistance. Although not every state's ability to manage the immediate aftermath of an accident will be equally dependent on international support, many countries will undoubtedly not be able to cope well, if at all, without foreign assistance. Certainly, this was a key lesson of the Chernobyl accident, which prompted the international community to adopt and bring into force—in record time—the Assistance Convention. In the context of the Fukushima accident, the Convention was not invoked; nor was RANET, the Convention's operational response and assistance network,[189] utilized.[190] Instead, Japan did receive direct assistance from other states, international governmental and industry organizations, and others,[191] which enabled the country to manage an emergency that clearly challenged national response capabilities.[192] The Fukushima experience does not, of course, detract from the crucial importance of the Convention-based international assistance regime for the management of nuclear or radiological emergencies. To the contrary, the accident provided valuable lessons[193] and, as noted, has led to an expansion of IAEA's "assessment and prognosis" mandate,[194] thereby ensuring the Agency an even more prominent role in assisting states in emergency situations in the future.

---

[189] RANET has been defined as "a network of States Parties to the Assistance Convention that are capable and willing to provide, upon request, specialized assistance by appropriately trained, equipped and qualified personnel with the ability to respond in a timely and effective manner to nuclear or radiological incidents and emergencies." See IAEA (2018).

[190] See IAEA, Technical Volume 3/5, *supra* note 14, at 96.

[191] For details see ibid. at 157–158.

[192] See The Fukushima Daiichi Accident, *supra* note 2, at 96.

[193] See IAEA, Technical Volume 3/5, *supra* note 14, at 157.

[194] See *supra* text at notes 2159–162 and See IAEA (2015), IAEA (2019) and (2020).

Although there is substantial treaty practice, overwhelmingly of a bilateral nature,[195] which covers mutual assistance in the event of a nuclear emergency, the centerpiece of the international assistance framework is undoubtedly the Assistance Convention and its mechanisms as discussed earlier.[196] As noted, states parties to the Convention are expected, within the limits of their capabilities, to identify national assistance capabilities (NACs) consisting of experts, equipment, and materials that could be made available to assist another state party.[197] This requirement is being met by registering NAC with RANET.[198] Registration does not, however, imply a legal commitment to assist. Nor does the Convention itself, of course, impose such an obligation. Rather, when a request is directed to the IAEA, it is the Agency's IEC that, upon receiving the request for international assistance, will attempt to co-ordinate and match registered NAC with the requesting state's specific needs through consultations with the offering state(s). The type of assistance that can be provided in this manner includes nuclear installation assessment and advice; radiation surveying and environmental sampling; dose assessment; decontamination; and medical support.[199]

To date, only 35 states have registered their capabilities with RANET (IAEA Press Release, 2020). Some countries may point to the Assistance Convention's lack of an implementation review process similar to that of the nuclear safety conventions as evidence of the allegedly "voluntary basis" of the international assistance network and blame the latter for states parties' lack of motivation to participate in RANET.[200] Whatever the real reasons for their reluctance, states' limited participation in RANET remains a matter of concern, especially as there is a growing need for national expertise to assist the Agency with accident assessment and prognosis and for the creation of "national rapid response teams that could be made available internationally through RANET."[201] Conversely, the IAEA Fukushima Report draws attention to the fact that a requesting state's failure to

---

[195] A relatively rare example of a multilateral agreement is the Nordic Mutual Emergency Agreement in Connection with Radiation Accident, IAEA Doc. INFCIRC/49, November 8, 1963. Mention must be made here also, of course, of the all-purpose EU Civil Protection Mechanism, first established in 2001, which covers also nuclear EPR actions. It is available for response assistance intervention throughout the EU as well as outside. See Decision No 1313/2013/EU of the European Parliament and the Council of December 17, 2013 on a Union Civil Protection Mechanism, OJ L 347/924, December 20, 2013.

[196] See *supra* text following note 32.

[197] Article 2, Paragraph 4 of the Convention.

[198] The European Emergency Response Centre that plays a similar role within the framework of the EU Civil Protection Mechanism could act as the "single registration point" in Europe for IAEA's RANET system. See ENCO Report, *supra* note 64, at 192.

[199] For details, see, e.g., IAEA, Response and Assistance Network, *supra* note 189, at 36−67.

[200] See Russian Proposal for Strengthening and Implementation of the Convention on Early Notification of a Nuclear Accident and the Convention on Assistance with the (sic) Case of Nuclear Accident or Radiological Emergency, Non-Paper, April 20, 2012, Sixth Meeting of the Representatives of Competent Authorities, *supra* note 147, at 49.

[201] IAEA Action Plan, *supra* note 15, at 3.

make in advance arrangements for receiving emergency assistance may prevent the state from being able to accept international assistance in the early stages of the national response."[202] The revised IAEA General Safety Requirements therefore now expressly call upon governments to "ensure that adequate arrangements are in place to [allow the country to] benefit from … international assistance for preparedness and response for a nuclear or radiological emergency."[203]

5. Conclusions

While, notionally, nuclear EPR today is still a national responsibility, there is no denying the fact that in the wake of the Fukushima accident EPR is increasingly being "internationalized." Clearly, the March 11, 2011, accident was a wake-up call for the international community that improvements in the international regulatory regime were needed not only with regard to the operational safety of nuclear power plants worldwide but also in respect of the management of off-site consequences of accidents through EPR. For Fukushima and its aftermath brought into sharp relief, the fact that effective emergency planning is a critical factor shaping public perceptions of nuclear safety and hence public acceptance of nuclear power generation, not just locally or nationally, but globally.

In drawing on the lessons of Fukushima, the IAEA, regional organizations, especially the European Union, the industry, and other stakeholders have, understandably, been pushing hard for the harmonization of national EPR measures and policies as a prerequisite for boosting the credibility of nuclear EPR generally. Significant efforts have been spent on improving the development and distribution of relevant nuclear emergency information, including on effective communication with target audiences, in particular the public, in other words on meeting the "the one message—many voices" challenge.[204] The net result of these efforts has been a deepening and tightening of bilateral, cross-border information sharing arrangements, better integration of regional preparedness and response measures, and, under the aegis of IAEA's IEC, the emergence of an effective global emergency communication, evaluation, and response system.

Today, it is safe to say, the global nuclear safety regime, including EPR, is significantly stronger and more resilient. Nevertheless, some challenges remain to be tackled. Foremost among these is perhaps the need to expand upon and make internationally mandatory, periodic in-country and on-site peer reviews of national emergency plans. Additionally, states need to bolster the Agency's incident and

---

[202] The Fukushima Daiichi Accident, *supra* note 2, at 96.

[203] GSR Part 7(IAEA, 2015), at, 45.

[204] See, e.g., Report on International Symposium on Communicating Nuclear and Radiological Emergencies to the Public, October 1–5, 2018, Vienna, Austria, 66 (emphasizing the importance of proper "messaging" as part of any emergency communications to the public and the media).

emergency system, by signing up in greater numbers to RANET and registering more of the requisite specific assets in support of the Agency's expanded assessment and prognosis mandate.

References

Adede, A.O., 1987. The IAEA Notification and Assistance Conventions in Case of a Nuclear Accident: Landmarks in the Multilateral Treaty-Making Process.

Advanced Nuclear Power Reactors, February 2020, Available from: https://www.world-nuclear.org/information-library/nuclear-fuel-cycle/nuclear-power-reactors/advanced-nuclear-power-reactors.aspx (Accessed 2, February 2020).

ANSN, 2013. Asian Nuclear Safety Network (ANSN) Progress Report 2013, pp. 36—40.

Armonización de los criterios reguladores para países de la región iberoamericana en la preparación y respuesta a emergencias radiológicas y nucleares, Available from: http://www.foroiberam.org/areas-colaborativas/preparacion-y-respuesta-a-emergencias (Accessed 12, September 2016).

ASEANTOM, 2013. Summary of 1st Meeting of ASEAN Network of Regulatory Bodies on Atomic Energy (ASEANTOM), Phuket, Thailand, September 3—4, 2013.

Chernobyl's disastrous cover-up is a warning for the next nuclear age. The Guardian, April 4, 2019.

Commission Implementing Decision of July 12, 2019 on the Annual Action Program 2019 for Nuclear Safety Cooperation to be financed from the general budget of the Union, Doc. C(2019) 5169 final, Annex 1, 5, Available from: https://webgate.ec.europa.eu/multisite/devco/sites/devco/files/action_document_for_nuclear_safety_cooperation_2019.pdf (Accessed 16, February 2020).

Conference Report, 2015. International Conference on Global Emergency Preparedness and Response. , October 19—23, 2015, Vienna, Austria, 12.

Council Conclusions on Off-site Nuclear Emergency Preparedness and Response, Doc. 14618/15, December 15, 2015, Annex, 2, 6.

Council of the European Union, September 17, 2013. Report on the Implementation of the Obligations under the Convention on Nuclear Safety—6th Review Meeting of the Contracting Parties. Doc. 13691/13, 42 (discussing the EURDEP's use on a global level by the IAEA).

Council of the European Union Council Conclusions on Off-site Nuclear Emergency Preparedness and Response, Doc. 14618/15, December 15, 2015, Annex 4.

Durand-Poudret, 2015. Toward a New International Framework for Nuclear Safety: Developments from Fukushima to Vienna. 95 Nuclear Law Bulletin 27 at 38—39.

Economic Research Institute for ASEAN and East Asia, April 17, 2015. The 2nd WG Meeting on 'Study for Building a Guideline and a Cooperative Framework in East Asian Countries in Case of Radioactive Emergency. Available from: http://www.eria.org/news/FY2015/04/the-2nd-wg-meeting-on-study-for-building-a-guIbideline-and-a-cooperative-framework-in-east-asian-count.html. (Accessed 27 December 2019).

ENCO Report Review of Current Off-Site Nuclear Emergency Preparedness and Response Arrangements in EU Member States and Neighbouring Countries, ENCO, Final Report, Main Text, Doc. ENER/D1/2012-474 at 11, December 2013..

European Commission-funded study of the feasibility of enhancing regional cooperation within the Association of Southeast Asian Nations (ASEAN) on radiological and nuclear emergency preparedness and response, February 16, 2016.

European Nuclear Safety Regulators Group, 2012. Peer Review Report: Stress Tests Performed on European Nuclear Power Plants.

Fifteen Years after the Chernobyl Accident: Lessons Learned, International Conference, Kiew, Ukraine, April 18−20, 2001, Executive Summary 3−5.

French and Agryris, May 2014. Nuclear Emergency Management: driven by precedent or international guidance?. In: Proceedings of the 11th International ISCRAM Conference— University Park, Pennsylvania, USA.

Gen IV International Forum, 2013 Annual Report, Appendix 1, 115.

Goldberg, S.M., Rosner, R., 2011. Nuclear Reactors: Generation to Generation.

Handl, G., 2016. Nuclear Off-Site Emergency Preparedness and Response: Some International Legal Aspects. In: Black-Branch, J.L., Fleck, D. (Eds.), Nuclear Non-Proliferation in International Law-Volume III: Legal Aspects of the Use of Nuclear Energy for Peaceful Purpose 337−339.

Handl, G., 1988. Après Tchernobyl: Quelques réflexions sur le programme législatif multilateral à l'ordre du jour, 92 RGDIP 5, at 58−59.

Handl, G., 1992. Grenzüberschreitendes nukleares Risiko und völkerrechtlicher Schutzanspruch 74−91.

Handl, 2003. The IAEA Nuclear Safety Conventions: An Example of Successful 'Treaty Management'? 72 Nuclear Law Bulletin, 7.

Harvey S., Visualising Data for Emergency Response: IAEA Launches International Radiation Mapping System, Available from: https://www.iaea.org/newscenter/news/visualising-data-for-emergency-response-iaea-launches-international-radiation-mapping-system (Accessed 3, February 2020).

HERCA, 2013. Practical Proposals for Further Harmonisation of the Reactions in European Countries to any Distant Nuclear or Radiological Emergency 5. Available from: https://www.herca.org/docstats/HERCA-WGE%20distant%20accidents%20recommendations%20report.pdf. (Accessed 25 February 2020).

HERCA, November 2015. Guidance for Bilateral Arrangements, p. 5.

HERCA, 2014. Emergency Preparedness: HERCA-Approach for a Better Cross-Border Coordination of Protective Actions During the Response in the Early Phase of a Nuclear Accident. development and practical testing 15. Available from: http://www.herca.org/docstats/HERCA_Approach_on_emergencies.pdf. (Accessed 27 December 2019).

HERCA, Emergency Preparedness and Response, Available from: http://www.herca.org/activities.asp?p=3&s=6 (Accessed 25, February 2020).

HERCA-WENRA Approach for a Better Cross-Border Coordination of Protective Actions during the Early Phase of a Nuclear Accident, October 22, 2014, 7 Available from: http://www.herca.org/uploaditems/documents/HERCA-WENRA%20approach%20for%20better%20cross-border%20coordination%20of%20protective%20actions%20during%20the%20early%20phase%20of%20a%20nuclear%20accIbide.pdf (Accessed 26, January 2020).

Higginbotham, A., 2019. Midnight in Chernobyl: The Untold Story of the World's Greatest Nuclear Disaster 169−179.

Holo, E., NKS Fukushima Seminar, January 12−13, 2016, Available from, 2016. Regional Cooperation on Nuclear and Radiological EPR. : http://www.nks.org/en/news/nks-fukushima-seminar-12-13-january-2016-in-stockholm.htm. (Accessed 27 October 2019).

IAEA Preparedness and Response for a Nuclear or Radiological Emergency, IAEA Safety Requirements, No. GS-R-2, 37 (2002).

IAEA Action Plan on Nuclear Safety, three to four, Available from: https://www.iaea.org/sites/default/files/actionplan-ns.pdf, (Accessed 19 October, 2016).

IAEA, Developing a regional emergency response plan in the Gulf region with the IAEA's technical cooperation support, Available from: https://www.iaea.org/technicalcooperation/Home/Highlights-Archive/Archive-2013/08302013_GCC_Emergency_Response.html (Accessed 12, September 2016).

IAEA, 2007. IAEA Safety Glossary: Terminology Used in Nuclear Safety and Radiation Protection. IAEA: 2007 Edition, Vienna 68–69.

IAEA, 2014. Nuclear Safety Review 2014, Doc. GC(58)/INF/3, 21.

IAEA Network for Emergency Assistance Grows to 35 Countries as India Joins," IAEA Press Release, January 7, 2020, Available from: https://www.iaea.org/newscenter/pressreleases/iaea-network-for-emergency-assistance-grows-to-35-countries-as-india-joins (Accessed 21, January 2020).

IAEA Report on Assessment and Prognosis in Response to a Nuclear or Radiological Emergency, International Experts Meeting, April 20–24, 2015, Vienna, Austria, 20.

IAEA Safety Standards, 2007. Arrangements for Preparedness for a Nuclear or Radiological Emergency. Safety Guide No. GS-G-2.1, 11.

IAEA, June 1996. Defence in Depth in Nuclear Safety. Report by the International Nuclear Safety Advisory Group, Doc. INSAG-10.

IAEA, 2003. Method for Developing Arrangements for Response to a Nuclear or Radiological Emergency. Doc. EPR-METHOD (2003), p. 42.

IAEA, 2007. Arrangements for Preparedness for a Nuclear or Radiological Emergency. IAEA Safety Guide, No. GS-G-2.1.

IAEA, 2009. Severe Accident Management Programmes for Nuclear Power Plants. Safety Guide No. NS-G-2.15,1.

IAEA, 2011. Criteria for Use in Preparedness and Response for a Nuclear or Radiological Emergency. IAEA Safety Guide, No. GSG-2.

IAEA, August 27–31, 2012. 2nd Extraordinary Meeting of the Contracting Parties to the Convention on Nuclear Safety. Final Summary Report. IAEA Doc. CNS/ExM/2012/04/Rev.2, para. 17.

IAEA, July 4, 2012. Sixth Meeting of the Representatives of Competent Authorities Identified under the Convention on Early Notification of a Nuclear Accident and the Convention on Assistance in Case of a Nuclear Accident or a Radiological Emergency. IAEA Doc. CAM/REP/2012, TM-41005.

IAEA, 2013. Actions to Protect the Public in an Emergency due to Severe Conditions at a Light Water Reactor. Doc. EPR-NPP Public Protective Actions 2013, 129.

IAEA, 2013. Integrated Regulatory Review Service (IRRS): Guidelines for the Preparation and Conduct of IRRS Missions. Foreword and 7–9.

IAEA, December 18, 2014. Joint Convention on the Safety of Spent Fuel Management and on the Safety of Radioactive Waste Management: Guidelines regarding the Form and Structure of National Reports. IAEA Doc. INFCIRC/604/Rev.3.

IAEA, 2014. Radiation Protection and Safety of Radiation Sources: International Basic Safety Standards. General Safety Requirements Part 3, No. GSR Part 3, 415.

IAEA, 2014. Radiation Protection and Safety of Radiation Sources: International Basic Safety Standards. IAEA General Safety Requirements Part 3, No. GSR Part 3.

IAEA, January 16, 2015. Guidelines regarding National Reports under the Convention on Nuclear Safety. IAEA Doc. INFCIRC/572/Rev.5.

IAEA, 2015. Report on Assessment and Prognosis in Response to a Nuclear or Radiological Emergency, International Experts Meeting, April 20−24, 2015, Vienna, Austria, p. 31.

IAEA, 2015. IAEA Report on International Experts Meeting: Severe Accident Management in the Light of the Accident at the Fukushima Daiichi Nuclear Power Plant. March 17−20, 2014, Vienna, Austria.

IAEA, 2015. Preparedness and Response for a Nuclear or Radiological Emergency. IAEA Safety Standards—General Safety Requirements, No. GSR Part 7.

IAEA, 2015. The Fukushima Daiichi Accident, Technical Volume 3/5: Emergency Planning and Response.

IAEA, October 3, 2016. Eighth Meeting of the Representatives of Competent Authorities Identified under the Convention on Early Notification of a Nuclear Accident and the Convention on Assistance in the Case of a Nuclear Accident or Radiological Emergency. Doc. CAM/REP/2016, TM-51,831.

IAEA, 2016. See Long Term Structure of the IAEA Safety Standards and Current Status, February 4, 2016. Available from: http://www-ns.iaea.org/committees/files/CSS/205/status.pdf. (Accessed 15 October 2016).

IAEA, Nuclear Safety Review 2016, Doc. GC (60)/INF/5, Paragraph 205.

IAEA, 2017. Report of the Technical Meeting on Next Generation Reactors and Emergency Preparedness and Response. held in IAEA, Vienna, between February 13 and 17, 2017.

IAEA IAEA Response and Assistance Network, Doc. EPR−RANET, August 1, 2018, Foreword.

IAEA, 2018. Emergency Preparedness Review (EPREV) Guidelines 1.

IAEA, 2018. Response and Assistance Network. Doc. EPR−RANET 2018, 13.

IAEA, 2019. Nuclear Safety Review 2019. Doc. GC(63)/INF/3, Paragraph 263.

IAEA, 2020. Operations Manual for IAEA Assessment and Prognosis during a Nuclear or Radiological Emergency. Doc. EPR-A&P (2019), pp. 20−22.

IAEA, Incident and Emergency Center, Available from: http://www-ns.iaea.org/tech-areas/emergency/incIbident-emergency-centre.asp?s=1 (Accessed 29, February 2020).

IAEA, 2020. Operations Manual for Incident and Emergency Communication. Doc. EPR−IEComm (2019).

IAEA, March 14, 2014. Experts' Meeting to Discuss Severe Accident Management after Fukushima. Available from: http://www-pub.iaea.org/iaeameetings/cn233Presentations.aspx. (Accessed 12 November 2019).

IAEA, 2015. Measures to Strengthen International Cooperation in Nuclear, Radiation, Transport and Waste Safety. IAEA Doc. GC(59)/RES/9, September 2015, Paragraph 31.

IAEA, September 16, 2019. IAEA and ASEAN Strengthen Cooperation in Nuclear Science, Technology and Applications, and Nuclear Safety, Security and Safeguards. Available from: https://www.iaea.org/newscenter/news/iaea-and-asean-strengthen-cooperation-in-nuclear-science-technology-and-applications-and-nuclear-safety-security-and-safeguards.

IAEA, 2020. International Radiological Information Exchange (IRIX) Format. Doc. EPR-IEComm (2019) Attachment 3, 4.

IAEA, 2020. International Radiation Monitoring Information System. Doc. EPR-IEComm 2019, Attachment 2, 4.

Institute of Nuclear Power Operations, August 2012. Lessons Learned from the Accident at the Fukushima Daiichi Nuclear Power Station, INPO 11−1005 Addendum.

International Experts' Meeting on Severe Accident Management in the Light of the Accident at the Fukushima Daiichi Nuclear Power Plant, Vienna, Austria March 17−20, 2014, Chairman's Summary, Available from: http://www-pub.iaea.org/iaeameetings/46832/International-Experts-Meeting-on-Severe-AccIbident-Management-in-the-Light-of-the-AccIbident-at-the-Fukushima-Daiichi-Nuclear-Power-Plant, (Accessed 12 November, 2019).

International Federation of Red Cross and Red Crescent Societies, Nuclear and Radiological Emergency Guidelines: Preparedness, Response and Recovery (2015).

Jayakumar, S., Koh, T., Beckman, R., Duy Phan, H. (Eds.), 2015. Transboundary Pollution: Evolving Issues of International Law and Policy 190, at 209−219.

Joint Radiation Emergency Management Plan of the International Organizations, Doc. EPR—JPLAN (2017).

Kovan, July 31, 2005. NEA's role in radiological protection−Keeping things real. Nuclear News.

Kubanyi, J., et al., 2008. Risk Informed Support of Decision Making in Nuclear Power Plant Emergency Zoning: Generic Framework toward Harmonizing NPP Emergency Planning Practices. JRC Scientific and Technical Reports 21−24.

Lang, W., 1988. Frühwarnung bei Nuklearunfällen. 39 Österr. Zeitschrift f. öffentliches Recht 9.

Mahjoub, A., 2015. AAEA Role in Improving EPR Coordination Interventions among Arab Countries. Presentation at the International Conference on Global Emergency Preparedness and Response, October 19−23, 2015, Vienna, Austria, Available from: https://nucleus.iaea.org/sites/iec/epr-conference-2015-docs/default.aspx. (Accessed 12 September 2016).

McMahon, J., 2011. Why Does FDA Tolerate More Radiation than EPA? FORBES, April 14, 2011, Available from: http://www.forbes.com/sites/jeffmcmahon/2011/04/14/why-does-fda-tolerate-more-radiation-than-epa/#4f5567027679. (Accessed 21 March 2017).

Measures to Strengthen International Cooperation in Nuclear, Radiation, Transport and Waste Safety, IAEA Doc. GC(55)/RES/9, September 2011, para. 82.

Meschenmoser, P., September 17, 2015. IAEA Launches Self-Assessment Tool for Emergency Preparedness. Available from: https://www.iaea.org/newscenter/news/iaea-launches-self-assessment-tool-emergency-preparedness. (Accessed 27 October 2016).

National Research Council, 2014. Lessons Learned from the Fukushima Nuclear Accident for Improving Safety of U.S. Nuclear Plants.

National Research Council, 2014. Lessons Learned from the Fukushima Nuclear Accident for Improving Safety of U.S. Nuclear Plants 216.

NEA/CNRA/CSNI, 2014. Joint Workshop on Challenges and Enhancements to Defense in Depth (DID) in Light of the Fukushima Daiichi NPP Accident. Workshop Proceedings, Paris, 5 June 2013, Doc. NEA/CNRA/R(2014) 4, p. 11.

NEA/OECD, 2002. Chernobyl: Assessment of Radiological And Health Impacts—2002 Update of Chernobyl: Ten Years On 121−127.

IAEA says Fukushima fallout warrants more evacuation. New Scientist, March 31, 2011. Available from: http://www.newscientist.com/article/dn20324-iaea-says-fukushima-fallout-warrants-more-evacuation.html#.U8BCsbEo4dV. (Accessed 14 March 2014).

Nordic Intervention Criteria for Nuclear or Radiological Emergencies − Recommendations The Radiation Protection Authorities in Denmark, Finland, Iceland, Norway and Sweden, 2001.

Nuclear disaster fallout 'would be no worse than living in London. The Times, November 23, 2017.

Nuclear Regulatory Commission, August 9, 2019. Mitigation Beyond-Design-Basis Events, 84 Federal Register 39,684.

Nuclear Transparency Watch, March 8–11, 2015. Position Paper of NTW on Emergency Preparedness & Response (EP&R) Situation in Europe.

OECD/NEA, 2006. International Nuclear Law in the post-Chernobyl Period 7. at 9–13.

OECD/NEA, 2016. Five Years after the Fukushima Daiichi Accident: Nuclear Safety Improvements and Lessons Learnt 26.

OECD/NEA, 2016. Implementation of Defense in Depth at Nuclear power Plants: Lessons Learned from the Fukushima Daiichi Accident 15.

OECD/NEA, 2018. Experience from the Fifth International Nuclear Emergency Exercise (INEX-5): Notification, Communication and Interfaces Related to Catastrophic Events Involving Ionising Radiation and/or Radioactive Materials.

Office for Nuclear Regulation, September 2011. Japanese Earthquake and Tsunami: Implications for the UK Nuclear Industry. Final Report, HM Chief Inspector of Nuclear Installations, pp. 144–145.

Protective Measures in Early and Intermediate Phases of A Nuclear or Radiological Emergency: Nordic Guidelines and Recommendations, 2014.

Reinberger, D., Ajanovic, A., Haas, R., 2019. The Technological Development of Different Generations and Reactor Concepts. In: Haas, R., Mez, L., Ajanovic, A. (Eds.), The Technological and Economic Future of Nuclear Power 243, pp. 249–252.

Report of the European Nuclear Safety Regulator's Group (ENSREG), November 25, 2015.

6 th Review Meeting of the Contracting Parties to the Convention on Nuclear Safety, March 24–April 4, 2014, Vienna, Austria, Summary Report IAEA Doc. CNS/6RM/2014/11_Final, Paragraphs 28–29.

Study on Good Practices in Implementing the Requirements on Public Information in the Event of an Emergency, under the EURATOM Basic Safety Standards Directive and Nuclear Safety Directive, Final Report (2019) [hereafter: Study on good practices] 13, Available from: https://op.europa.eu/en/publication-detail/-/publication/8bf9a38e-2531-11ea-af81-01aa75ed71a1 (Accessed 26, January 2020, emphasizing the need for early cross-border information exchanges at the preparedness stage).

The Fukushima Daiichi Accident, Report by the Director General, IAEA Doc. GC(59)/14, at 4 (2015).

The Guardian, June 3, 2018. What was the fallout from Fukushima? The Guardian. Available from: https://www.theguardian.com/environment/2018/jun/03/was-fallout-from-fukushima-exaggerated. (Accessed 21 January 2020).

Thomas, P.J., 2017. Quantitative guidance on how best to respond to a big nuclear accident. 112 Process Safety and Environmental Protection 4.

Tomić, B., July 1–12, 2018. Need for Off-site Nuclear EP & R. Training Course on Nuclear Off-site Emergency Preparedness, Singapore.

U.S. Nuclear Regulatory Commission, 2011. Recommendations for Enhancing Reactor Safety in the 21st Century: The Near-Term Task Force Review of Insights from the Fukushima Dai-Ichi Accident.

U.S. urges Americans within 50 miles of Japanese nuclear plant to evacuate; NRC chief outlines dangerous situation. The Washington Post, March 16, 2011. Available from: http://www.washingtonpost.com/national/us-urges-americans-within-50-miles-of-japanese-nuclear-plant-to-evacuate/2011/03/16/ABwTmha_story.html. (Accessed 21 September 2016).

Vienna Declaration on Nuclear Safety on Principles for the Implementation of the Objective of the Convention on Nuclear Safety to Prevent Accidents and Mitigate Radiological Consequences, Doc. CNS/DC/2015/2/Rev.1, Annex 1, February 9, 2015.

19 inside WANO No. 3, 4 WANO after Fukushima: Strengthening Global Nuclear Safety, 2011.

Western European Nuclear Regulators' Association (WENRA), September 24, 2014. Report on Safety Reference Levels for Existing Reactors: Update in Relation to Lessons Learned from Tepco Fukushima Dai-ichi Accident, pp. 33–35.

Yutthana, T., July 11–12, 2018. Nuclear EPR in the ASEAN Region. Training Course on Nuclear Emergency Preparedness and Response. ESI/CIL, National University of Singapore. Available from: http://esi.nus.edu.sg/docs/default-source/doc/session-2-2-tumnoi-yutthana-nuclear-epr-in-the-asean-region.pdf?sfvrsn=2. (Accessed 23 January 2020).

International conventions and legal frameworks on nuclear safety, security, and safeguards

Paul Murphy[1], Ira Martina Drupady[2]

[1]*Managing Director, Murphy Energy & Infrastructure Consulting, LLC, Washington, DC, United States;* [2]*Research Associate, Energy Studies Institute, National University of Singapore, Singapore*

LIST OF ABBREVIATIONS

3S	safety, security, and safeguards
CBM	Confidence-building measure
CD	UN Conference on Disarmament
CIS	Commonwealth of Independent States
CNS	Convention on Nuclear Safety
CPPNM	Convention on the Physical Protection of Nuclear Material
CSA	Comprehensive safeguards agreement
CTBT	Comprehensive Nuclear Test-Ban Treaty
ECA	Export credit agency
EIA	Environmental impact assessment
Euratom	European Atomic Energy Community
FMCT	Fissile Material Cut-off Treaty
HEU	Highly enriched uranium
IAEA	International Atomic Energy Agency
IFC	International Finance Corporation
IMS	International Monitoring System
INIR	Integrated Nuclear Infrastructure Review
INPO	Institute of Nuclear Power Operators
IRRS	Integrated Regulatory Review Service
MEA	Multilateral environmental agreement
MOP	Meeting of Parties
NNWS	Non−nuclear-weapon states
NPP	Nuclear power project
NPT	Treaty on the Non-Proliferation of Nuclear Weapons
NUS	National University of Singapore
NWS	Nuclear-weapon states
OECD	Organisation for Economic Co-operation and Development
OSART	Operational Safety Review Team
OSI	On-site inspection
PRTR	Pollutant release and transfer register

Advanced Security and Safeguarding in the Nuclear Power Industry. https://doi.org/10.1016/B978-0-12-818256-7.00007-6

TNT	Trinitrotoluene
UN	United Nations
UNECE	UN Economic Commission for Europe
WANO	World Association of Nuclear Operators

1. Part 1: what is meant by nuclear safety, security, and safeguards?

The nuclear sector is unique in several respects:

- First, the sector is bifurcated by both military (i.e., weapons) and civilian (i.e., energy, medical applications, etc.) uses.

  Military considerations arise from the use of atomic weapons at the end of World War II, followed by the proliferation of such weapons thereafter among the larger powers (the United States, the Soviet Union, the United Kingdom, France, and China, which, not coincidentally, are the five permanent members of the United Nations' Security Council). During the post–World War II era, known as the Cold War, the increase in nuclear weapons led to the deterrence theory of "mutually assured destruction."

  The threat of nuclear devastation has greatly influenced how the nuclear sector has been viewed, particularly the misplaced association of nuclear weapons with civilian nuclear applications, principally power generation. Nevertheless, given the potential devastation of nuclear weapons, considerable efforts have been made to limit the spread of nuclear weapons, including a focus on the technology or "know-how" necessary to create nuclear weapons.

  This convergence of military and civilian applications is best observed in the discussion of the safety, security, and safeguards regimes, which combine to create separation of civilian uses from weaponization of the technology and other concerns regarding potential radiological releases and nonproliferation issues.

- Second, the nuclear sector has an international organization — the International Atomic Energy Agency ("IAEA"), which is an instrumentality of the United Nations ("UN") — that plays roles that match the dual nature of the sector.

  The IAEA has oversight and enforcement functions that are embedded within international treaty structures, principally for treaty adherence. In this role, the IAEA has specific authority to act. Contrast that authority with the role played by the IAEA in the civilian nuclear sector, where the IAEA has an extensive portfolio of guidelines and standards that are available for Member States to use, as they develop their civilian nuclear programs. While such guidelines and standards are not binding (i.e., they do not have legal effect), they carry tremendous weight within the nuclear industry and with the wider stakeholder audience (e.g., governments, lenders, investors, regulatory authorities, contractors, suppliers/vendors, nongovernmental organizations, and the general public). As a result, such guidelines and standards are very influential for the nuclear industry, giving the IAEA significant authority.

In the power industry, such authority and guidance roles are unmatched, recognizing that no institution akin to the IAEA exists for oil, coal, gas, hydro, or renewables.

The IAEA's prominence only exists because of the cooperative nature of the nuclear industry. This cooperation spans two key ideas: cooperation and information exchange. These ideas were embedded in United States' President Dwight D. Eisenhower's "Atoms for Peace" speech in 1953, which led to the establishment of the IAEA. Cooperation is necessary to achieve the collective goals of safety, security, and safeguards, noting the interdependent nature of such goals (as will be discussed in the following).

Cooperation is also important to harness the beneficial power of the atom, recognizing that such benefits have to be transferred from a small number of countries that possess the technology and know-how for civilian use to those countries that wish to share in such benefits, while acknowledging that such information sharing and technological exchange must be done responsibly (i.e., in a manner that does not divert such knowledge and/or material to military purposes). Information sharing also promotes industry best practices. Not only is the IAEA a source for information sharing, but also other industry organizations, such as the World Association of Nuclear Operators ("WANO") and the Institute of Nuclear Power Operators ("INPO"), serve to share lessons learned and other best practices to promote a robust and responsible civilian nuclear industry.

This cooperation extends to crisis situations, whereby Member States have committed to exchange information and to provide assistance in the event of a radiological event. Such cooperation and assistance falls under the category, "Emergency Preparedness and Response." This category of activity recognizes the potential severity of a radiological event, the potential cross-border impact, and the larger impact on the environment. Such activity considers both static facilities and nuclear material in transit. It is a system "designed to reduce the risk of emergencies and to mitigate their consequences." (IAEA, 2003, p. 75).

- Third, on the civilian nuclear side, for a nuclear power project to move forward, engagements must occur at both the national and the project levels for the project to be viable.

Given that only sovereign entities can enter into international and bilateral treaty commitments, and such treaty commitments are essential components of a responsible civilian nuclear program, project development requires alignment between national commitments and commercial considerations at the project level. Furthermore, any commitments in these areas also need to be reflected in the national law of a Member State, which puts a paramount importance on the alignment between a Member State's sovereign/international obligations and the implementation of such obligations at the national level, including the structure, function, and enforcement of such laws.

Coupled with the role of national regulatory authorities (see the following), this two-tiered relationship between state and project must be managed for successful project development.

- Fourth, countries with civilian nuclear programs have national nuclear regulatory authorities that maintain an oversight function over civilian nuclear programs. While other commercial sectors have national regulatory authorities with safety mandates (e.g., aviation), the impact of nuclear regulatory authorities is unique. This uniqueness is observed in how the nuclear regulatory process has a significant impact on both cost and schedule, with regulatory challenges being one of the major reasons why large-scale nuclear power projects often run over schedule and budget. While other industries seem to tolerate certain levels of death and severe injury (e.g., automobiles, aviation, mining, etc.), the nuclear industry has put such a premium on safety that the result has been, in effect, a "zero tolerance" policy. While such an approach has yielded the best safety record in the asset class, it does present particular challenges for the industry, particularly in project development.

These unique factors combine under the concepts of safety, security, and safeguards (sometimes referred to as "3S").

- "Safety" conceptually focuses on the safe operation of nuclear facilities and handling of nuclear materials. It is the idea that nuclear facilities must be operated, and nuclear materials must be handled, in a manner that protects life and property. The safety concept involves a recognition of the power of the atom (and the potential human impact of ionizing radiation).
  The IAEA's "Handbook on Nuclear Law" notes two subsidiary principles:
  - The "Prevention Principle" holds that, "given the special character of the risks of using nuclear energy, the primary objective of nuclear law is to promote the exercise of caution and foresight so as to prevent damage that might be caused by the use of the technology and to minimize any adverse effects resulting from misuse or from accidents" (IAEA, 2003, p. 6).
  - The "Protection Principle" holds that "[t]he fundamental purpose of any regulatory regime is to balance social risks and benefits. Where the risks associated with an activity are found to outweigh the benefits, priority must be given to protecting public health, safety, security and the environment. Of course, in the event that a balance cannot be achieved, the rules of nuclear law should require action favouring protection. It is in this context that the concept commonly referred to as the 'precautionary principle' (i.e., the concept of preventing foreseeable harm) should be understood" (IAEA, 2003, p. 6).
- "Security" conceptually focuses on physical protection. It involves two elements — the physical security of the nuclear facility and the security (i.e., control) of radiological sources/materials.
  Facility security considers both military and political aspects of nuclear facilities. Nuclear facilities can potentially be targets of both terrorists and anti-uclear

groups. Given the sensitivities around nuclear materials, it is of paramount importance to maintain both boundary security and the security of the nuclear materials themselves. Lost or abandoned materials can pose a health threat to the public, and certain types of nuclear materials could be used for terrorist acts and/or to support the creation of a nuclear device by either state or nonstate actors. The security concept extends to materials in transit.

- "Safeguards" conceptually focus on nonproliferation.
  Essentially, safeguards involve the prevention of nuclear material being diverted from peaceful use to the production of nuclear weapons or other radiological devices. As noted in the Handbook, "safeguards comprise three functions: accountancy, containment and surveillance, and inspection" (IAEA, 2003, p. 121).
  - Accountancy involves a Member State reporting the types and quantities of fissionable material under its control to the IAEA.
  - Containment and surveillance measures involve the IAEA's use of seals on nuclear material containers and filmed or televised recordings of key areas at nuclear facilities.
  - Inspections are performed by the IAEA to verify declared quantities of nuclear material and that no undeclared nuclear material exists in the Member State.

  Nonproliferation concepts are also embedded in national legal regimes, which address the export of nuclear technology and the potential diversion of certain technologies for military purposes (the case of "dual use" items, which could have both civilian and military applications).

In addition to the 3S concept, nuclear liability (legal responsibility for property damage and/or personal injury arising out of a radiological event at a nuclear facility) is another area where international conventions are prominent. This subject is addressed in Chapter 8 of this book. Without repeating what is addressed therein, nuclear liability forms the final pillar of the international structure around the nuclear industry. As will be discussed in the following, suitable international nuclear liability coverage becomes one of the threshold issues in nuclear power project development. Coverage for cross-border damage or injury in the case of a radiological event serves to bind the fellow treaty Member States, similar to the ways the 3S structure links together Member States in the areas of safety, security, and safeguards.

Taken together, the ideas of safety, security, safeguards, and nuclear liability create a structure and system that:

- recognizes the sensitive nature of nuclear technology, striking a balance between the potential benefits of civilian applications with the potential threats of military applications, particularly in the case of the technology falling into the hands of rogue state and nonstate actors;
- recognizes the need for Member States to cooperate in the exchange of nuclear information, materials, and technology;

- stresses the importance of public safety, including the safe operation of nuclear facilities and the custodianship of nuclear materials;
- creates points of responsibility and of oversight to ensure compliance with the 3S concept;
- promotes cooperation, support, and information exchange, particularly in serious cases involving a radiological release; and
- establishes a central authority for maintaining the structure, to include both oversight/enforcement and standards/guidance.

2. Part 2: what are the key international conventions on safety, security, and safeguards?

2.1 Introduction

Part 2 of this chapter reviews the major international treaties concerning "safety," "security," and "safeguards." In other scholarship, "Emergency Preparedness and Response" is often treated as a separate heading; however, the authors have elected to keep it within the "Safety" heading, so as to preserve the 3S concept throughout the discussion in this chapter.

In reviewing each of the treaties/conventions, it is important to recognize that such agreements only bind the Member States that are signatories thereto. In other words, to the extent that a country is not a signatory to the agreement, it is not bound by the agreement, which, understandably, creates certain gaps within these regimes.

Furthermore, in assessing which countries are parties to a treaty/convention, it is also important to recognize that mere signature is not sufficient for a treaty to be effective. For a country to be bound by its terms, the country must submit its instrument of ratification for its full participation. In addition, for the agreement to be enforceable within the legal/judicial structure of a country, further implementing legislation might have to be enacted, following ratification.

In addition, certain treaties/conventions have conditions before the treaty/convention takes effect. Traditionally, this involves a certain number of Member States having submitted their instruments of ratification, at which point the agreement will then have binding effect.

In the list that follows, some agreements have not become effective; nevertheless, they are described, given their significance and overall level of support.

2.2 Safety

2.2.1 Convention on nuclear safety

The Convention on Nuclear Safety ("CNS") was adopted on September 20, 1994, and came into force on October 24, 1996, with the IAEA Director General as Depositary. It currently has 88 Parties and 65 Signatories (IAEA, 2019e). The CNS is the first legally binding international treaty to address the safety of nuclear installations. Prior to its coming into force, nuclear safety was considered wholly a national legal matter. It was the 1986 Chernobyl accident that gave impetus to this Convention. The event raised political concerns worldwide, particularly in the context of the potential transboundary aspects of a major nuclear power plant accident. The breakup of the Soviet Union in 1991 further exacerbated global concerns regarding the safety of nuclear installations in Commonwealth of Independent States ("CIS").

The CNS governs the safety of nuclear installations, defined as land-based civil nuclear power plants, under a Contracting Party's jurisdiction (CNS, 1994, Article 2). These include storage, handling, and treatment facilities for radioactive materials located on the same site of the nuclear power plant, directly related to its operation. It seeks to ensure that such installations are operated in a safe, well-regulated, and environmentally sound manner. More specifically, the purpose of the CNS is to (CNS, 1994, Article 1):

- achieve and maintain a high level of nuclear safety worldwide through the enhancement of national measures and international cooperation including, where appropriate, safety-related technical cooperation;
- establish and maintain effective defenses in nuclear installations against potential radiological hazards to protect individuals, society, and the environment from harmful effects of ionizing radiation from such installations; and
- prevent accidents with radiological consequences and mitigate such consequences should they occur.

Each Contracting Party of the CNS is required to take, within the framework of its national law, the legislative, regulatory and administrative measures, and other steps necessary to implement its obligations under the Convention (CNS, 1994, Article 4). This includes, among others, establishing an independent regulatory body to implement the required legislative and regulatory framework on nuclear safety, with adequate authority, competence, and human and financial resources (CNS, 1994, Article 8); ensuring that on- and off-site emergency plans are in place, are routinely tested, and cover the activities to be carried out in the event of an emergency (CNS, 1994, Article 16[1]); and taking appropriate steps to ensure that the siting, design and construction, and operation of a nuclear installation are in accordance with its obligations under the Convention (CNS, 1994, Article 17—19). Each Contracting Party must also ensure that prime responsibility for the safety of a nuclear installation rests with the holder of the relevant license and take appropriate steps to ensure that each the license holder meets its responsibility (CNS, 1994, Article 9).

In the case of nuclear installations existing at the time the CNS comes into force, each Contracting Party is required to take appropriate steps to ensure that the safety of such installations is reviewed as soon as possible (CNS, 1994, Article 6). In this regard, when necessary, the Contracting Party is required to ensure that all reasonably practicable improvements to upgrade the safety of the nuclear installation are made as a matter of urgency. If such upgrading cannot be achieved, plans are to be put into place to shut down the facility as soon as practically possible.

The CNS is neither a regulatory regime (no inspections and no international secretariat) nor a sanctions regime (no penalties for noncompliance), and therefore, procedures for resolving disputes are vague. It provides that, in the event of a disagreement between two or more Contracting Parties concerning the interpretation or application of the Convention, Contracting Parties should consult within the framework of a meeting of the Contracting Parties with a view to resolving the disagreement (CNS, 1994, Article 29).

Rather, the CNS is considered to be an incentive convention with periodic peer review meetings held to encourage Contracting Parties to meet CNS obligations. Prior to each review meeting held, each Contracting Party is required to submit a report on measures it has taken to implement its obligations under the Convention (CNS, 1994, Articles 20 and 21).

2.2.2 Joint convention on the safety of spent fuel management and on the safety of radioactive waste management

The Joint Convention on the Safety of Spent Fuel Management and on the Safety of Radioactive Waste Management ("Joint Convention") was adopted on September 29, 1997, and came into force on June 18, 2001, with the IAEA Director General as Depositary. It currently has 82 Parties and 42 Signatories (IAEA, 2019g). It is the first legally binding international treaty committing participating States to a global nuclear regime that strives to achieve and maintain a high level of safety in the areas of spent fuel management and radioactive waste management.

The Joint Convention governs the safety of spent fuel management resulting from the operation of civilian nuclear reactors, the safety of radioactive waste management when the radioactive waste results from civilian application, and the safety of certain discharges (Joint Convention, 1997, Article 3). It does not apply to spent fuel or radioactive waste from military or defense programs.

The stated purpose of the Joint Convention is to (Joint Convention, 1997, Article 1):

- achieve and maintain a high level of safety worldwide through the enhancement of national measures and international cooperation;
- ensure that there are effective defenses against potential hazards to protect individuals, society, and the environment from harmful effects of ionizing radiation (so that the needs and aspirations of the present generation are met without compromising the ability of future generations to meet their needs and aspirations); and
- prevent accidents with radiological consequences and mitigate their consequences should they occur.

Contracting Parties of the Joint Convention are required to take the appropriate legislative, regulatory, and administrative measures to ensure that, at all stages of spent fuel management and radioactive waste management, individuals, society, and the environment are adequately protected against radiological hazards in the siting, design, construction, and assessment of facilities, as well as their operations and closure (Joint Convention, 1997, Article 4−17). This includes ensuring that, before and during operation of a spent fuel or radioactive waste management facility, there are appropriate on-site and, if necessary, off-site emergency plans. (Joint Convention, 1997, Article 25[1]). Each Contracting Party is also required to take the appropriate steps for the preparation and testing of emergency plans for its territory insofar as it is likely to be affected in the event of a radiological emergency at a spent fuel or radioactive waste management facility in the vicinity of its territory. (Joint Convention, 1997, Article 25[2]).

In terms of reporting, each Contracting Party is required to attend and be represented at meetings of the Contracting Parties (Joint Convention, 1997, Article 33 [1]). In each of these review meetings, the Contracting Party is required to submit a national report on measures it has taken to implement the obligations of the Joint Convention prior to each review meeting of the Contracting Parties (Joint Convention, 1997, Article 32).

As with the CNS, under the Joint Convention, rules for enforcement are vague. In the event of a disagreement between two or more Contracting Parties concerning the interpretation or application of the Convention, the Contracting Parties are required to consult within the framework of a meeting of the Contracting Parties with a view to resolving the disagreement (Joint Convention, 1997, Article 38). In the event that such consultations are ineffective, recourse can be made to the mediation, conciliation, and arbitration mechanisms provided for in international law, including the rules and practices prevailing within the IAEA.

2.2.3 Convention on early notification of a nuclear accident

The Convention on Early Notification of a Nuclear Accident ("Early Notification Convention") was adopted in Vienna on September 26, 1986, and in New York on October 6, 1986, and came into force on October 27, 1986, with the IAEA Director General as Depositary. It currently has 124 Parties and 69 Signatories (IAEA, 2019d). Under the Early Notification Convention, obligations apply to States that are party to the Convention. Private parties (e.g., operators and suppliers) do not fall under its purview.

The Early Notification Convention aims to strengthen the international response to nuclear accidents to minimize transboundary radiological consequences by providing a mechanism for rapid information exchange and a notification system for nuclear accidents. The Convention applies in the event of any accident involving specified facilities or activities of a State Party from which a release of radioactive material occurs or is likely to occur and which has resulted, or may result, in an international transboundary release that could be of radiological safety significance to another State (Early Notification Convention, 1986, Article 1).

In the event of an accident, the State Party must immediately notify States which are or may be physically affected (as well as the IAEA itself), of the accident, its nature, the time of its occurrence, and its exact location, where appropriate. (Early

Notification Convention, 1986, Article 2[a]). Moreover, it must promptly provide those States and the IAEA with available information relevant to minimizing the radiological consequences in those States (Early Notification Convention, 1986, Article 2[b]).

In this regard, each State Party is required to notify other States Parties and the IAEA of its competent authorities, point of contact, and a focal point responsible for issuing and receiving notification and information (Early Notification Convention, 1986, Article 7[1]). The IAEA is required maintain an up-to-date list of national authorities that serve as competent authorities and points of contact responsible for issuing and receiving notification and information related to notification of a nuclear accident and to provide such information to the States Parties (Early Notification Convention, 1986, Article 7[3]).

Although there are provisions in the Early Notification Convention for the settlement of disputes, there are no provisions dealing with enforceability or setting forth consequences for noncompliance. Disputes are to be settled first by negotiation or any other peaceful means of dispute resolution acceptable to the disputing parties (Early Notification Convention, 1986, Article 11). If a dispute cannot be settled by negotiation, matters are referred to arbitration or to the International Court of Justice ("ICJ") for decision.

2.2.4 Convention on assistance in the case of a nuclear accident or radiological emergency

The Convention on Assistance in the Case of a Nuclear Accident or Radiological Emergency ("Assistance Convention") was adopted in Vienna on September 26, 1986, and in New York on October 6, 1986, and came into force on February 26, 1987, with the IAEA Director General as Depositary. It currently has 119 Parties and 68 Signatories (IAEA, 2019c). Under the Convention, obligations apply to the States that are party to the Convention. Private parties (e.g., operators and suppliers) do not fall under its purview.

The Assistance Convention aims to strengthen the international response to a nuclear accident or radiological emergency, including those that result from terrorist or other malicious acts. In the event of any accident involving facilities or activities of a State Party or persons or entities under its control, States Parties are required to cooperate among themselves and with the IAEA to facilitate prompt requests and provision of assistance (Assistance Convention, 1986, Article 1). These arrangements, be it at the bilateral or multilateral levels or a combination of both, are intended to mitigate the consequences of such an accident; protect life, property, and the environment from the effects of radioactive releases; and promote, facilitate, and support cooperation among States Parties.

A State Party may call for assistance from any other State Party, and each State Party to which a request for such assistance is directed may decide and notify the requesting State Party, directly or through the IAEA, whether it is in a position to render the assistance requested; as well as the scope and terms of the assistance that might be rendered. (Assistance Convention, 1986, Article 2[3]). In this regard, the requesting party is required to provide local facilities and services for the proper and effective administration of the assistance. It should also ensure the protection of personnel, equipment, and materials brought into its territory by or on behalf of the assisting party for such purposes. (Assistance Convention, 1986, Article 3[b]). Unless otherwise agreed, it is also to reimburse the assisting party for costs incurred for the services and for all expenses in connection with the assistance. (Assistance Convention, 1986, Article 7[2]). The requesting State is also to afford personnel of the assisting party and personnel acting on its behalf the necessary privileges, immunities, and facilities for the performance of their assistance functions (Assistance Convention, 1986, Article 8[1]), which is important vis-à-vis nuclear liability.

Similar to the Early Notification Convention, the Assistance Convention requires the State Party to notify the IAEA and other States Parties of its competent authorities and point of contact authorized to make and receive requests for, and to accept offers of, assistance (Assistance Convention, 1986, Article 4[1]). The IAEA is also required to maintain an up-to-date list of these national authorities and provide such information to the States Parties (Assistance Convention, 1986, Article 4[3]).

Under the Assistance Convention, in the case of claims, States Parties are obligated to cooperate closely to facilitate the settlement of legal proceedings and claims (Assistance Convention, 1986, Article 10). The immunity and indemnification of a party providing assistance extends only to damages caused within the jurisdiction of a state requesting assistance. Accordingly, transboundary damages would not be covered under the Assistance Convention and would not be indemnified.

Disputes are to be settled first by negotiation or any other peaceful means of dispute resolution acceptable to the disputing parties. (Assistance Convention, 1986, Article 13[1]). If a dispute cannot be settled by negotiation, matters are referred to arbitration or to the ICJ for decision (Assistance Convention, 1986, Article 13[2]).

2.3 Security

2.3.1 Convention on the physical protection of nuclear material

The Convention on the Physical Protection of Nuclear Material ("CPPNM") was adopted on October 26, 1979, and came into force on February 8, 1987, with the IAEA as Depositary. It currently has 160 Parties and 68 Signatories (IAEA, 2019f). It is one of the 13 counterterrorism instruments and is the only internationally legally binding undertaking in the area of the physical protection of nuclear material.

The CPPNM governs nuclear material used for peaceful purposes while in international nuclear transport and, with certain exceptions, while in domestic use, storage, and transport. In this regard, it facilitates the safe transfer of nuclear material and the physical protection of nuclear material in domestic use, storage, and transport and criminalizes, within the national law of the States Parties, the violation of the protection of nuclear material.

Under the CPPNM, States Parties are required to undertake several key obligations for the physical protection of nuclear material. Among others, a State Party is required to take appropriate steps to ensure that, during international nuclear transport, nuclear material within its territory or on board a ship or aircraft under its jurisdiction that is engaged in the transport to or from that State is protected (CPPNM, 1979, Article 3). Each State Party is also required not to import, export, or authorize the import or export or the transit of nuclear material unless the State Party has received assurances that such material will be protected during its international transport (CPPNM, 1979, Article 4). States Parties are also required to identify and make known their central authority and point of contact having responsibility for physical protection of nuclear material and for coordinating recovery and response operations in the event of a credible threat or an actual occurrence of any unauthorized removal, use, or alteration of nuclear material (CPPNM, 1979, Article 5[1]).

In the case of a credible threat or an actual occurrence of theft, robbery, or any other unlawful taking of nuclear material, States Parties are required to, in accordance with their national law, provide cooperation and assistance to the maximum feasible extent in the recovery and protection of such material to any State that so requests such aid (CPPNM, 1979, Article 5[2]). States Parties are also required to cooperate and consult regarding guidance on the design, maintenance, and improvement of systems of physical protection for nuclear material in international transport (CPPNM, 1979, Article 5[3]).

In terms of reporting requirements of the CPPNM, each State Party must inform the depositary of its laws and regulations that give effect to the Convention. (CPPNM, 1979, Article 14[1]). The State Party where an alleged offender is prosecuted is required to communicate the final outcome of the proceedings to the States directly concerned and then to the depositary who shall inform the States (CPPNM, 1979, Article 14[2]). However, the CPPNM neither has inspection requirements nor remedies for noncompliance.

In the event of a dispute between two or more States Parties concerning the interpretation or application of the Convention, they are required to consult with a view to the settlement of the dispute by negotiation, or by any other peaceful means of settling disputes acceptable to all parties to the dispute (CPPNM, 1979, Article 17 [1]). Any dispute of this character that cannot be settled shall, at the request of any party to such dispute, be submitted to arbitration or referred to the ICJ for decision (CPPNM, 1979, Article 17[2]). Where a dispute is submitted to arbitration, if, within 6 months from the date of the request, the parties to the dispute are unable to agree on the organization of the arbitration, a party may request the President of the ICJ or the UN Secretary-General to appoint one or more arbitrators (CPPNM, 1979, Article 17[2]).

2.3.2 Amendment to the convention on the physical protection of nuclear material

In July 2005, the States Parties agreed to amend the CPPNM to strengthen its provisions and reduce the vulnerability of States Parties to nuclear terrorism. The

Amendment to the CPPNM ("Amendment") was adopted on July 8, 2005, and came into force on May 8, 2016. It currently has 123 Parties (IAEA, 2019b). Upon its entry into force, the Amendment is also thereafter known as the Convention on the Physical Protection of Nuclear Material and Nuclear Facilities. However, the IAEA Secretariat has elected to continue to refer to the "CPPNM" and to the "Amendment to the CPPNM" until all States Parties to the CPPNM have consented to be bound by the Amendment, in line with established depositary practice.

The purpose of the Amendment is to achieve and maintain worldwide effective physical protection of nuclear material used for peaceful purposes and of nuclear facilities used for peaceful purposes; to prevent and combat offenses relating to such material and facilities worldwide; and to facilitate cooperation among State Parties (Amendment, 2016, Article 1A). It broadens the scope of the physical protection of nuclear material and of nuclear facilities used for peaceful purposes. Whereas the obligations for physical protection under the CPPNM only cover nuclear material during international transport, the Amendment also covers nuclear facilities, nuclear material in domestic use, and storage and transport used for peaceful purposes, as well as sabotage (Amendment, 2016, Article 1[d and e]). It explicitly excludes "activities of armed forces during an armed conflict" and "activities undertaken by military forces in the exercise of their official duties" from the scope of the Convention, inasmuch as they are covered by other rules of international law (Amendment, 2016, Article 2). The Amendment also explicitly excludes nuclear material used or retained for military purposes and nuclear facilities containing such material (Amendment, 2016, Article 2).

The Amendment requires States to establish, implement, and maintain a physical protection regime applicable to nuclear material and facilities under their jurisdiction, including an appropriate legislative and regulatory framework for physical protection; a competent authority responsible for its implementation; and other administrative measures necessary for the physical protection of such material and facilities (Amendment, 2016, Article 2A). It also requires States to bring under their jurisdiction and make punishable under their national laws, certain offences, including theft, robbery, smuggling of nuclear material, or sabotage of nuclear facilities, as well as acts related to directing and contributing to the commission of such offences (Amendment, 2016, Article 7). The Amendment also strengthens cooperation, assistance, and coordination between States and the IAEA, including introducing new arrangements for exchange of information (Amendment, 2016, Article 5).

2.3.3 International convention for the suppression of acts of nuclear terrorism

The International Convention for the Suppression of Acts of Nuclear Terrorism ("Nuclear Terrorism Convention") was adopted on September 14, 2005, and came into force on July 7, 2007, with the UN Secretary-General as Depositary. It currently has 116 Parties and 115 Signatories (UN Treaty Collection, 2020c). The Convention is one of 13 antiterrorism instruments and the first one since the September 11, 2001, attacks. Designed as a preemptive instrument, it aims to establish a reliable international legal mechanism to enhance international cooperation among States at all stages of combating nuclear terrorism.

The Nuclear Terrorism Convention criminalizes the planning, threatening, or carrying out acts of nuclear terrorism. Thus, it requires State Parties to criminalize these offenses through national legislation and to establish penalties in line with the gravity of the offences (Nuclear Terrorism Convention, 2005, Articles 5 and 6). It also requires States Parties to cooperate in preventing or prosecuting acts of nuclear terrorism (Nuclear Terrorism Convention, 2005, Article 7). This includes taking all practicable measures to prevent (and to undertake counterpreparations in respect of) nuclear terrorism offences in their respective territories that would then be committed within or outside their territories (Nuclear Terrorism Convention, 2005, Article 7[1a]); exchanging accurate and verified information in accordance with their national law (Nuclear Terrorism Convention, 2005, Article 7[1b]); and informing the UN Secretary-General of their competent authorities and liaison points responsible for sending and receiving the information (Nuclear Terrorism Convention, 2005, Article 7[4]). The Convention also sets out conditions under which States may establish jurisdiction for offenses (Nuclear Terrorism Convention, 2005, Article 9) and guidelines for extradition and other measures of punishment (Nuclear Terrorism Convention, 2005, Articles 9–17).

In terms of reporting, the State Party where the alleged offender is prosecuted shall, in accordance with its national law or applicable procedures, communicate the final outcome of the proceedings to the UN Secretary-General, who shall then transmit the information to other States Parties (Nuclear Terrorism Convention, 2005, Article 19). However, there are neither inspections nor remedies for noncompliance.

2.4 Safeguards

2.4.1 The treaty on the non-proliferation of nuclear weapons

The Treaty on the Non-Proliferation of Nuclear Weapons ("NPT") was adopted on July 1, 1968 and came into force on March 5, 1970, with the Russian Federation, the United Kingdom of Great Britain and Northern Ireland, and the United States as Depositary Governments. It currently has 191 Parties and 93 Signatories (UNODA, n.d.).

The NPT is a "grand bargain" between the five nuclear-weapon states ("NWS") and the non–nuclear-weapon states ("NNWS") across the three pillars of nonproliferation, disarmament, and the peaceful use of nuclear energy. The Treaty aims to prevent the wider dissemination of nuclear weapons; facilitate the application of IAEA safeguards on peaceful nuclear activities; support the effective safeguarding of the flow of source and special fissionable materials by use of instruments and other techniques at certain strategic points; cease the nuclear arms race at the earliest possible date and undertake effective measures in the direction of nuclear disarmament; seek to achieve the discontinuance of all test explosions of nuclear weapons for all time and continue negotiations to this end; and cease the manufacture of nuclear weapons, liquidate all existing stockpiles, and eliminate national arsenals of nuclear weapons and the means of their delivery.

Under the NPT, an NWS is defined as one that has manufactured and exploded a nuclear weapon or other types of nuclear explosive devices prior to January 1, 1967 (NPT, 1970, Article 9). An NWS Party to the Treaty undertakes not to transfer nuclear weapons or other nuclear explosive devices or control over such weapons or explosive devices, whether directly, or indirectly, to any recipient whatsoever (NPT, 1970, Article 1). It is also required not in any way to assist, encourage, or induce any NNWS to manufacture or otherwise acquire nuclear weapons or other nuclear explosive devices, or have control over such weapons or explosive devices (NPT, 1970, Article 1).

An NNWS Party to the Treaty undertakes not to receive the transfer of nuclear weapons or other nuclear explosive devices or control over such weapons or explosive devices, whether directly, or indirectly, from any transferor (NPT, 1970, Article 2). It is also required not to manufacture or otherwise acquire nuclear weapons or other nuclear explosive devices and not to seek or receive any assistance in the manufacture of nuclear weapons or other nuclear explosive devices (NPT, 1970, Article 2). An NNWS is also required to accept safeguards, as set forth in an agreement to be negotiated and concluded with the IAEA in accordance with the Statute of the IAEA and the Agency's safeguards system, for the exclusive purpose of verification of the fulfillment of its obligations assumed under the NPT (NPT, 1970, Article 3). More details of what an IAEA Safeguards Agreement entails are discussed in the following section.

In terms of disarmament obligations, all States Parties to the Treaty undertake to pursue negotiations in good faith on effective measures relating to the cessation of the nuclear arms race at an early date and to nuclear disarmament, and to pursue a treaty on general and complete disarmament under strict and effective international control (NPT, 1970, Article 6).

All States Parties to the NPT undertake to facilitate and have the right to participate in the fullest possible exchange of equipment, materials, and scientific and technological information for the peaceful uses of nuclear energy (NPT, 1970, Article 4[1]). Those States in a position to do so are required cooperate in contributing alone or together with other States or international organizations to the further development of the applications of nuclear energy for peaceful purposes, especially in the territories of NNWS Parties to the NPT, with due consideration for the needs of the developing areas of the world (NPT, 1970, Article 4[2]).

There are neither reporting nor inspection requirements under the NPT. However, while there are no direct remedies for noncompliance, political or economic sanctions by other parties could arise separately.

2.4.2 IAEA safeguards agreements

As mentioned, under the NPT, each NNWS Party to the Treaty is also required to accept IAEA safeguards that are embedded in legal-binding agreements negotiated and concluded with the Agency for the exclusive purpose of verification of the fulfillment of its obligations assumed under the Treaty (IAEA, n.d.[b]). In this regard, the IAEA Statute provides the Agency with the authority to establish and administer safeguards (IAEA Statute, 1989, Article 3). When the IAEA Board of Governors approves a safeguards agreement, it authorizes the Director General to conclude and subsequently implement it.

Most safeguards agreements are comprehensive safeguards agreements ("CSAs") that have been concluded by the IAEA with NNWS Parties to the NPT and nuclear weapon–free zone treaties. Under a CSA, the IAEA has the right and obligation to ensure that safeguards are applied on all nuclear material in the territory, jurisdiction, or control of the State for the exclusive purpose of verifying that such material is not diverted to nuclear weapons or other nuclear explosive devices. To date, the IAEA has concluded CSAs with 175 States and some 100 of these States have also concluded small quantities protocols to their CSAs (IAEA, n.d.[b]). The small quantities protocol is a protocol that may be concluded in conjunction with a CSA. The original small quantities protocol was made available to States with minimal or no nuclear material and no nuclear material in a "facility."

In addition, the five NWS Parties to the NPT have concluded voluntary offer safeguards agreements with the IAEA. Under a voluntary offer agreement, the IAEA applies safeguards to nuclear material in facilities that the State has voluntarily offered and the IAEA has selected for the application of safeguards. Under the agreement, the IAEA verifies that nuclear material remains in peaceful activities and is not withdrawn from safeguards, except as provided for in the agreement.

Safeguards are also implemented in three States not party to the NPT − India, Pakistan, and Israe − lon the basis of item-specific agreements they have concluded with the IAEA. Under these agreements, the IAEA applies safeguards to ensure that nuclear material, facilities, and other items specified under the safeguards agreement are not used for the manufacture of any nuclear weapon or to further any military purpose and that such items are used exclusively for peaceful purposes and not for the manufacture of any nuclear explosive device.

2.4.3 Model protocol additional to the agreements(s) between states(s) and the international atomic energy agency for the application of safeguards

The Model Protocol Additional to the Agreement(s) between State(s) and the International Atomic Energy Agency for the Application of Safeguards ("Additional Protocol") contributes to global nuclear nonproliferation objectives by strengthening the effectiveness of the safeguards system and improving its efficiency. It is not a stand-alone agreement, but rather a protocol to a safeguards agreement that grants the IAEA expanded rights of access to information and locations in the States, thereby providing the IAEA additional tools for verification. To date, Additional Protocols are in force in 136 States and the European Atomic Energy Community ("Euratom") (IAEA, 2019a). Another 15 States have signed an Additional Protocol but have yet to bring it into force, whereas one State provisionally applies an Additional Protocol pending its CSA's entry into force.

The Additional Protocol is designed for States with any type of safeguards agreement with the IAEA. A State with CSAs that decides to conclude an Additional Protocol must accept all provisions of the Model Protocol Additional document, whereas States with item-specific or voluntary offer agreements may accept and implement those measures of the Model Additional Protocol that they are prepared to accept.

For States with CSAs, concluding an Additional Protocol helps fill the gaps in the information reported under a CSA through such measures as: broader access to information about the State's nuclear program, increased physical access by the IAEA, and improved administrative arrangements. In this regard, an Additional Protocol would significantly increase the IAEA's ability to verify the peaceful use of all nuclear material in a particular State and, therefore, provide a much more robust picture of such State's nuclear programs, plans, nuclear material holdings, and trade.

2.4.4 Comprehensive nuclear test-ban treaty and protocol to the comprehensive nuclear test-ban treaty

The Comprehensive Nuclear Test-ban Treaty ("CTBT") was adopted on September 24, 1996; however, it is not yet in force. It will enter into force 180 days after the date of deposit of the instruments of ratification by all 44 States listed in Annex 2 to the Treaty, to the UN Secretary-General as Depositary (CTBT, 1996, Article 14). Of the 44 States included in Annex 2, all have signed with the exceptions of the Democratic People's Republic of Korea, India, and Pakistan (CTBT Preparatory Commission, n.d.). With the exception of these nonsignatory states, all Annex 2 States have signed and ratified the CTBT with the exceptions of China, Egypt, Iran, Israel, and the United States (CTBT Preparatory Commission, n.d.).

The CTBT establishes a de facto international norm to ban any nuclear weapon test explosion or any other nuclear explosion. Among other things, it causes each State Party to undertake not to carry out, cause, encourage, or in any way participate in any nuclear weapon test explosion or any other nuclear explosion and to prohibit and prevent any such nuclear explosion at any place under its jurisdiction or control.

The CTBT, together with the draft Fissile Material Cut-off Treaty ("FMCT"), is the integral component of the global nuclear control regime, providing the foundation for eventual nuclear disarmament. The FMCT is a proposed international agreement that would prohibit the production of the two main components of nuclear weapons: highly enriched uranium ("HEU") and plutonium. The proposed Treaty is discussed under the framework of the UN Conference on Disarmament ("CD"), established in 1979 as the sole multilateral negotiating forum on disarmament. States that are Party to the NPT as NNWS are already prohibited from producing or acquiring fissile material for weapons. An FMCT would extend restrictions to the five NWS that are Party to the NPT (i.e., United States, Russia, United Kingdom, France, and China), as well as the four that are not (i.e., Israel, India, Pakistan, and North Korea). It is unclear if the scope of the Treaty will include preexisting stocks of fissile material. States in favor of including preexisting stocks tend to call for a

Fissile Material Treaty (FMT), whereas States favoring a ban on production often refer to an FMCT.

The CTBT itself includes a Protocol in three parts: (1) detailing the International Monitoring System ("IMS"); (2) on-site inspections ("OSI"); and (3) confidence-building measures ("CBMs"). There are also two annexes to the Protocol: annex 1 detailing the location of various Treaty monitoring assets associated with the IMS and annex 2 detailing the parameters for screening events.

In parallel to the opening of the CTBT for signature, the Preparatory Commission for the Comprehensive Nuclear-Test-Ban Treaty Organization (the "Preparatory Commission") was set up in 1996 as an interim organization tasked with the establishment of the CTBT's verification regime and the promotion of signatures and ratifications of the Treaty, so that it can enter into force.

When the Treaty does enter into force, a State Party to the Treaty will be obligated, among other things, to establish and be a member of the Comprehensive Nuclear Test-Ban Treaty Organization ("Organization"), which is set out to achieve the object and purpose of the Treaty; to ensure the implementation of its provisions, including those for international verification of compliance with it; and to provide a forum for consultation and cooperation among State Parties. (CTBT, 1996, Article 2[1]). The Preparatory Commission will cease to exist.

When the Treaty is in force, a State Party is also required to cooperate with the Organization in the exercise of its functions by consulting on any matter, which may be raised relating to the object and purpose, or implementation of the provisions, of the Treaty. (CTBT, 1996, Article 2[5]). In this regard, a State Party is required to inform the Organization of the measures taken, in accordance with its constitutional processes, to (1) prohibit natural and legal persons anywhere on its territory or in any other place under its jurisdiction as recognized by international law from undertaking any activity prohibited to a State Party under the Treaty; (2) prohibit natural and legal persons from undertaking any such activity anywhere under its control; and (3) prohibit, in conformity with international law, natural persons possessing its nationality from undertaking any such activity anywhere. (CTBT, 1996, Article 3[1]). A State Party is also required to inform the Organization of satisfaction of its obligation to designate or set up a National Authority (CTBT, 1996, Article 3[4]).

The CTBT's verification regime consists of the following elements: (1) an IMS; (2) consultation and clarification; (3) OSIs; and (4) CBMs. (CTBT, 1996, Article 4 [1]). When the Treaty enters into force, a State Party undertakes to cooperate with the Organization and other States Parties to facilitate the verification of compliance to the Treaty by (1) establishing the necessary facilities to participate in verification measures and establish the necessary communication; (2) provide data obtained from national stations that are part of the IMS; (3) participate, as appropriate, in a consultation and clarification process; (4) permit the conduct of OSIs; and (5) participate, as appropriate, in CBMs (CTBT, 1996, Article 4[3]).

The State Party also undertakes to cooperate with the Organization and with other States that are party to the Treaty in the improvement of the verification regime and in the examination of the verification potential of additional monitoring technologies (CTBT, 1996, Article 4[11]). This is with a view to developing, when appropriate, specific measures to enhance the efficient and cost-effective verification of the Treaty. A State Party also undertakes to promote cooperation among other States that are party to the Treaty. (CTBT, 1996, Article 4[12]). This is to facilitate, and to participate to the fullest possible extent, the exchanges relating to technologies used in the verification of the Treaty, to enable all States Party to the Treaty to strengthen their national implementation of verification measures, and to benefit from the application of such technologies for peaceful purposes.

In terms of reporting requirements, once the CTBT is in force, the State Party is required to provide the Technical Secretariat with notification of any chemical explosion using 300 tons or greater of trinitrotoluene ("TNT") — equivalent blasting material detonated as a single explosion anywhere on its territory or at any place under its jurisdiction or control (CTBT Protocol, 1996, Part 3[2]). A Review Conference is to be held every 10 years after the Treaty's entry into force to reassess its operation and effectiveness and to ascertain that its objectives and purposes are being upheld (CTBT, 1996, Article 8).

The CTBT provides for measures to redress a violation of the Treaty and to ensure compliance, including sanctions, and for settlement of disputes. If the Conference or Executive Council determines that a case is of particular gravity, it can bring the issue to the attention of the UN (CTBT, 1996, Article 5).

---

3. Part 3: what are the additional conventions of relevance?

Part 3 of this chapter reviews two additional major treaties that are relevant for cooperation, consultation, information exchange, and public engagement, specifically in the context of transboundary environmental impact assessments ("EIA") for proposed nuclear energy—related activities. These are the Convention on Environmental Impact Assessment in a Transboundary Context ("Espoo Convention") and the Convention on Access to Information, Public Participation in Decision-Making and Access to Justice in Environmental Matters ("Aarhus Convention").

Although both Conventions originated as multilateral environmental agreements ("MEAs") negotiated under the auspices of the UN Economic Commission for Europe ("UNECE"), their provisions are not inherently regional (i.e., limited to European countries). Together, the Espoo and Aarhus Conventions provide a framework for transboundary EIAs and public participation that is far more comprehensive than can be found in specific nuclear law instruments comprising the global regime.

On a more practical level, the majority of nuclear power plants under construction are located in non-UNECE Member States (IAEA, 2019h, p. 10). As are many of the States currently seriously contemplating nuclear energy. The need for robust tools to help prevent and mitigate the potential adverse transboundary environmental impact of new nuclear power projects ("NPPs") and promote public participation will become increasingly important. The potential applicability of both Conventions to serve as reference points, or as actual legal instruments in other regions, is something worth contemplating.

A broad summary of each instrument is undertaken, including key oversight and enforcement provisions. The discussion subsequently highlights some of the key challenges posed by both instruments during different stages of a transboundary EIA for a proposed NPP and the public participant requirements it would entail. Parts of the analysis for this section were drawn from an unpublished report produced by the National University of Singapore ("NUS") under the ESI-CIL Nuclear Governance Project Phase 1 (2016—18).

3.1 Convention of environmental impact assessment in a transboundary context (the "Espoo Convention")

The Espoo Convention was adopted on February 25, 1991, and came into force on September 10, 1997, with the UN Secretary-General performing the functions of Depositary. As one of the five MEAs negotiated under the auspices of the UNECE, all 45 Parties (with the exception of the European Union) and 30 Signatories of the Convention are one of the 56 Member States of the UNECE (UN Treaty Collection, 2020b). Notably, a first amendment was adopted on February 27, 2001, to allow all UN Member States to accede to the Convention upon approval of the Meeting of Parties ("MOP") (Espoo Convention, 1991, Article 17). The amendment came into force on August 26, 2014; however, it has yet to become operational, pending 13 ratifications (UN Treaty Collection, 2020b).

The impetus for the Espoo Convention dates back to the 1972 Conference on the Human Environment that asserted the responsibility of States to ensure activities within their jurisdiction or control do not cause damage to the environment of other States or of areas beyond the limits of national jurisdiction (Stockholm Declaration, 1972, Principle 21). In this regard, the Espoo Convention aims to enhance international cooperation in assessing environmental impact, particularly in a transboundary context. It sets out obligations for Parties to assess the environmental impact of certain activities at an early stage of planning and lays down the general obligations for them to notify and consult each other on all major projects under consideration likely to have an adverse transboundary environmental impact.

More specifically, the Convention obliges Parties to take all appropriate and effective measures to prevent, reduce, and control significant adverse transboundary environmental impacts from proposed activities (Espoo Convention, 1991, Article 2 [1]). This includes undertaking the necessary legal, administrative, or other measures to implement provisions of the Convention, including with respect to proposed activities listed in Appendix 1 of the Convention likely to cause significant adverse transboundary impact; the establishment of an EIA procedure that permits public participation; and preparation of the EIA documentation described in Appendix 2 of the Convention (Espoo Convention, 1991, Article 2[2]). Parties of the Convention are also required to ensure that an EIA is undertaken prior to a decision to authorize or undertake a proposed activity that is likely to cause significant adverse transboundary impact (Espoo Convention, 1991, Article 2[3]) and that affected Parties are notified (Espoo Convention, 1991, Article 2[4]).

Under the Espoo Convention, if a dispute arises between two or more Parties about the interpretation or application of the Convention, they are required to seek a solution by negotiation or by any other method of dispute settlement acceptable to the parties to the dispute (Espoo Convention, 1991, Article 15[1]). For unresolved disputes, one or both of the following means of dispute settlement are accepted as compulsory in relation to any Party accepting the same obligations, namely, (1) submission of the dispute to the ICJ and (2) arbitration in accordance with the procedure set out in Appendix VII.3 of the Convention (Espoo Convention, 1991, Article 15[2]). If the parties to the dispute have accepted both means of dispute settlement referred to the earlier, the dispute may be submitted only to the ICJ, unless the parties agree otherwise (Espoo Convention, 1991, Article 15[3]).

In the context of nuclear energy—related activities, nuclear power reactors — regardless of their thermal output (small research reactors are excluded) — are listed in Appendix 1 of the Espoo Convention. An NPP proposed in the jurisdiction of a State Party of the Espoo Convention would therefore be within the scope of the Convention. However, a determination would still need to be made as to whether it is likely to cause significant adverse transboundary impact.

It was not until the fifth MOP held in June 2011 that steps were undertaken to begin applying the Convention to nuclear energy—related activities. The seventh MOP held in June 2017 endorsed the Good Practice Recommendations on the Application of the Convention to Nuclear Energy-related Activities ("Good Practice Recommendations"). Even so, there was a recognition that the document is not a legal interpretation of, nor imposes obligations under, the Convention.

More specifically under the Espoo Convention, the lack of internationally standardized procedures for determining whether a proposed nuclear energy—related activity is likely to cause significant adverse transboundary impact results in uncertainty as to when the Espoo Convention's requirement for an EIA applies. Although public participation is considered as an integral part of a transboundary EIA, it is unclear how affected Parties should be identified. Moreover, whereas the Convention contemplates public participation in two distinct phases, it is unclear when these Parties should be notified (Espoo Convention, 1991, Article 3[8]) and when translations of notification documentation are to be provided (Espoo Convention, 1991, Article 4[2]).

Furthermore, the Espoo Convention does not expressly provide for the scoping of the EIA documentation in consultation with all relevant stakeholders prior to its actual preparation, nor does it contemplate different levels of EIA that may be required. The requirement to include a description of alternatives in EIA documentation is only "where appropriate" (Espoo Convention, 1991, Appendix II[b]), suggesting that the decision be left to the Party of origin.

The Good Practices Recommendations suggest the EIA documentation process identify and assess all impacts of a nuclear energy activity throughout the whole life cycle, taking into consideration its impacts on climates as well as risks (UNECE,

2017, Paragraph 47). However, noticeably missing from this document is the definition of "impact." The Convention itself in its prescription of EIA documentation does not include an assessment of cumulative impacts of the proposed activities. The Espoo Convention and the Good Practices Recommendations also differ in prescription regarding the translation of EIA documentation or any parts of it into a nontechnical summary, which is a key element for providing information to the public.

3.2 Convention on access to information, public participation in decision-making and access to justice in environmental matters (the "Aarhus Convention")

The Aarhus Convention was adopted in June 25, 1998, and came into force on October 30, 2001. It currently has 47 Parties and 39 Signatories, with the UN Secretary General acting as Depositary (UN Treaty Collection, 2020a). Although the Convention is also one of the five MEAs negotiated under the auspices of the UNECE, any UN Member State may accede to it upon approval by the MOP (Aarhus Convention, 1998, Article 19[3]).

Beyond environmental concerns, the Aarhus Convention is about government accountability, transparency, and responsiveness. It links environmental rights with human rights and governmental accountability with environmental protection and focuses on interactions between the public and public authorities in a democratic context. In this regard, the "public" that has participatory rights in a decision-making process represents a potentially larger pool of participants than that contemplated under the Espoo Convention.

Together with its Kyiv Protocol on Pollutant Release and Transfer Registers ("Kyiv Protocol") that came into force on October 8, 2009, they are the only legally binding instruments on environmental democracy. Moreover, the Kyiv Protocol is the only legally binding international instrument on pollutant release and transfer registers ("PRTRs"). These are inventories of pollution from industrial sites and other sources that enhance public access to information through the establishment of coherent, nationwide PRTRs.

The Parties to the Aarhus Convention are required to make the necessary provisions so that public authorities (at national, regional, or other levels) will contribute to the aforementioned rights becoming effective (Aarhus Convention, 1998, Article 2). This means Parties are to take the necessary legal, administrative, and other measures to implement the provisions of the Convention, including measures to achieve compatibility among the provisions implementing the information, public participation, and access-to-justice provisions, as well as proper enforcement measures (Aarhus Convention, 1998, Article 3[1]). Unlike the Espoo Convention, however, the Aarhus Convention is not focused specifically on the transboundary EIA context, and therefore, there are no provisions prescribing the content of EIAs. Rather, the focus is on specific proposed activities (where the scope of proposed activities is substantially similar to that of the Espoo Convention and, in fact, identical in relation to nuclear-related installations). When a specific activity is within its scope, the Aarhus Convention establishes the rights of the public with regard to the environment in the areas of access to information, public participation, and access to justice.

Specifically, the Convention secures the rights of the public concerned to have due account taken of their views early on (when all options are open) in an environmental decision-making process (which is wider than, but includes, the transboundary EIA process) (Aarhus Convention, 1998, Article 6[2]). It gives the public concerned the right to access all information relevant to the decision-making process available at the time of the public participation procedure and contemplates feedback through two main channels: in writing or at a public hearing (Aarhus Convention, 1998, Article 6). The Convention further secures the right to access justice by giving the public concerned the right to challenge the substantive or procedural legality of any decision, act, or omission within the context of public participation in decisions on specific activities (Aarhus Convention, 1998, Article 9[2]). Limitations to this right to access of information and grounds for refusal include those disclosures that would adversely affect international relations, national defense, or public security or the confidentiality of commercial and industrial information and intellectual property rights (Aarhus Convention, 1998, Articles 3 and 4). In the context of the settlement of disputes, the Aarhus Conventions provides similar provisions to the Espoo Convention (Aarhus Convention, 1998, Article 16).

The Aarhus Convention Compliance Committee was established to fulfill the requirement of Article 15 of the Convention (UNECE, n.d.). This optional arrangement of a nonconfrontational, nonjudicial, and consultative nature allows members of the public to communicate concerns about a Party's compliance directly to a committee of international legal experts empowered to examine the merits of the case. The Committee cannot issue binding decisions but rather makes recommendations to the full MOP. However, as MOPs occur infrequently, Parties typically try to comply with the Committee's recommendations. This unique compliance review mechanism can be triggered in five ways (UNECE, 2019, Paragraph 83):

1. members of the public may make communications concerning a Party's compliance with the Convention;
2. a Party may make a submission about compliance by another Party;
3. a Party may make a submission concerning its own compliance;
4. the secretariat may make a referral to the Committee; or
5. the Meeting of the Parties may request the Committee to examine a Party's compliance with the Convention.

Following calls over several years for more practical guidance on how to improve the implementation of the Convention's provisions on public participation in decision-making, the Task Force on Public Participation in Decision-making prepared the Maastricht Recommendations on Promoting Effective Public Participation in Decision-Making in Environmental Matters ("Maastricht Recommendations"), which were endorsed by the MOP in July 2014. Although the Maastricht Recommendations are neither legally binding nor exhaustive, they are based on existing good practice and are intended as a practical tool to share expertise and good practice and assist the implementation of the Convention.

Similar to the Espoo Convention, the process of taking the necessary steps to apply the Aarhus Convention to nuclear energy—related activities has also been slow. The Aarhus Convention and Nuclear ("ACN") Initiative was only established in 2008 (7 years after the Convention came into effect) to assess the implementation of the Convention in the nuclear domain in Europe. The fourth meeting of the Task Force on Public Participation in Decision-Making, held in March 2013, served as the final ACN Initiative roundtable, where participants examined key lessons learned and considered possible action to further increase public engagement in nuclear activities (UNECE, 2013).

Under the Aarhus Convention, Article 6 does govern public participation in decisions on specific activities, including activities listed in Annex 1 to the Convention, which includes nuclear power stations (Aarhus Convention, 1998, Article 6[1]). However, beyond the challenge of identifying the affected Party/Parties under the Espoo Convention, the notification of the "public concerned" under the Aarhus Convention poses additional organizational and logistical challenges. The more detailed requirements for providing "relevant information" to the "public concerned" under the Aarhus Convention also create additional disclosure challenges for proposed nuclear energy—related activities. Moreover, although the Convention identifies a public hearing as one of the two main channels for feedback, it is unclear when a public hearing should be conducted, merely stating that it should be held as appropriate (Aarhus Convention, 1998, Article 6[7]). Like the Espoo Convention, the Aarhus Convention is also silent on the need for translation, although the Maastricht Recommendations make clear that language issues should be addressed as appropriate (Maastricht Recommendations, 2015, Paragraph 63). Specifically, project developers may find themselves involved in a review of the substantive and/or procedural legality of a decision, act, or omission following a challenge by a member of the "public concerned."

4. Part 4: what are the significance of these regimes? — reputational risk analysis

The structure and function of the 3S international treaty regime, along with international nuclear liability conventions, has been summarized in Parts 2 and 3 of this chapter. Further history on these regimes can be found from a number of other sources, and this chapter is not meant to reproduce that analysis. Instead, this chapter builds on the preceding Parts 1—3 to consider how these regimes impact the development of NPPs. In particular, when considering the challenge of financing NPPs, these regimes serve as a critical part of the analysis; thus, it is important to understand why financiers — both debt (lenders) and equity (investors) — analyze the international commitments of the host country within which the NPP is to be developed, as well as the international commitments of the countries from which the technology and nuclear know-how is exported.

For lenders in any major infrastructure project, a fundamental question in the financing decision is "How am I going to get paid back?" (or, slightly rephrased, "When am I going to get paid back?"). Such a basic question should come as no surprise to any reader. However, equally important to major international commercial lending institutions and to export credit agencies ("ECAs") is the question, "Is this a good project?" The first question is more empirical in nature, as financial modelers can look at project costs, projected project revenues, and corporate balance sheets to reach a conclusion as to the financial viability of a project. The second question is much more subjective in nature and falls within the concept of "reputational risk."

Specific to the nuclear power industry, this subjective analysis — this determination of whether a prospective NPP is a good project — is a threshold issue that can determine whether or not the NPP is financeable. While NPPs are not unique as large-scale infrastructure projects that require the mobilization of multiple billions of dollars in capital, the civilian nuclear power industry is subject to a heightened level of scrutiny and sensitivity, given the unique and sensitive nature of nuclear power, the history of several high-profile nuclear incidents, and the international regimes that form an integral part of the global civilian nuclear power industry. As a result, a robust understanding of the scope of this analysis is critical for project developers, host governments, prospective financiers, and other stakeholders.

As a preliminary matter, it should be noted that, to date, no NPP has ever been project financed (whereby the lenders, under a nonrecourse/limited recourse structure, look solely to the revenues generated by the project company — a special purpose vehicle existing solely for the purpose of owning and operating (either directly or indirectly) the revenue-generating asset — to repay the debt, as well as provide a subordinated equity rate of return). Despite a lack of project finance history in this sector, it is nonetheless relevant to address NPP financing in the context of the project (as opposed to an assessment of a corporate balance sheet alone), considering that many project financing principles are applied to NPPs and that such principles heavily influence the thinking of prospective financiers to a project. The lenders will apply the same sort of rigor to the prospective NPP that they would have applied to a classic project finance structure, principally because of the inherent sensitivities (whether fair or not) surrounding anything nuclear and the expanded reputational risk analysis that will be applied to an NPP. Thus, in attempting to answer the "good project" question, the rigor of the project finance diligence process is useful. For NPPs in the context of financing, it still is about the project, even if it is a balance sheet deal or a deal that is backed by a sovereign guarantee.

4.1 Reputational risk

"Reputational risk" is a topic that can encompass a number of concepts. It is not unique to the nuclear power industry; however, because of the unique characteristics of, and issues facing, NPPs, the idea of "reputational risk" encompasses a wider range and more significant set of considerations. Of note, both commercial banks and ECAs have specific lending guidelines/policies for NPPs; some examples are listed in the following:

1. BNP Paribas' nuclear policy at:
   https://group.bnpparibas/uploads/file/csr_sector_policy_nuclear_power.pdf;
2. Société Générale's nuclear policy at:
   https://www.societegenerale.com/sites/default/files/2018/civil_nuclear_power_sector_policy.pdf; and
3. Export-Import Bank of the United States' nuclear policy at:
   https://www.exim.gov/policies/ex-im-bank-and-the-environment/international-environmental-and-social-guidelines#a-3.

Such guidelines must be satisfied for such institutions to provide financing for a prospective NPP. Broadly speaking, these guidelines and further project diligence undertaken by lending institutions, as supported by their external technical and legal advisors, cover a number of issues, which are discussed in the following.

4.1.1 International standards and practices

Lenders want to know that the prospective NPP will meet internationally recognized standards. Through guidance issued by international organizations such as the IAEA and WANO, a wealth of information is available, reflecting a level of openness and cooperation in the nuclear industry due to, and through the existence of, such supranational (IAEA) and industry (WANO) organizations.

This concept can be covered in financing documentation by the concept of "Prudent Industry Practice." An example of a definition of Prudent Industry Practice as developed by the authors in work on applicable projects is as follows.

"Prudent Industry Practice" means the standards, practices, methods, and procedures consistent with that degree of skill, diligence, judgment, prudence, and foresight that would ordinarily be expected from an international skilled and experienced owner, contractor, equipment manufacturer, or, as the case may be, operator, engaged in designing, engineering, constructing, developing, commissioning, repairing, refurbishing, operating, insuring, maintaining, and/or decommissioning a nuclear power plant, in each case taking into account and giving appropriate consideration to all applicable standards and guidelines and local conditions.

A similar concept will be "international best practice," but the use of "best" can create less flexibility and more subjectivity in a particular case; consequently, a robust "Prudent Industry Practice" definition should provide financiers with the necessary level of technical scrutiny. Regardless of the word choice, lenders will work with their external advisors in assessing technical aspects of the project to ensure the NPP complies with this requirement, especially given the sensitivities (both perceived and real) involved in an NPP.

4.1.2 International agreements

Unlike other members of its peer group — whether the power industry or the infrastructure industry more broadly — the nuclear power industry operates within an international treaty framework that covers a number of subject areas, involving commitments at a Member State level. Lenders will want to see that the host government for the NPP is a Member State for this recognized set of international treaty commitments, and similarly, the lenders will want to make continued compliance with such treaties a condition of the financing, which necessarily creates a certain disconnect between the borrower (the party subject to the financing covenants) and the host government (the party ultimately responsible for compliance under the treaty).

When comparing nuclear power with its peer group, this structure of international treaty commitments as a means of holding the nuclear power industry accountable to a set of uniform obligations is unparalleled. Moreover, the presence of an international body — the IAEA — which provides a pseudogovernance function for the industry and has no comparison with other forms of power generation or infrastructure. This set of rules, coupled with the presence of the IAEA, provides lenders with a benchmark for assessing the quality of potential NPPs and the commitment of the host government to international standards and practices.

4.1.3 Public acceptance and sustained government commitment

As a result of the nuclear incidents at Three Mile Island (1979), Chernobyl (1986), and Fukushima (2011), NPPs are surrounded by a heightened sensitivity by the general public. NPP risks and the safety case are often misunderstood, and project developers and host governments will need to work with the public to develop the necessary level of local support for the project. Such stakeholder engagement will be an ongoing activity for the developers, owners, and governments involved; however, such engagement will be critical at the earlier stages (e.g., prefeasibility study, site selection, preconstruction) of the project's life cycle. The negative consequences of public acceptance are underscored by the histories at Shoreham (Long Island, United States), Bataan (Philippines), Zwentendorf (Austria), Kudankulam 1 and 2 (India), the cancellation of Italy's civilian nuclear power program following Chernobyl (and the referendum in 2011 that blocked an attempt by Enel to restart the program), and Germany's actions following Fukushima.

Government support is critical to the successful development of NPPs. This support manifests in several fashions: (1) financial support, both from the exporting country (as applicable) and the host country (for both the NPP and the overall civilian nuclear program); (2) consistent and sustained support in the legal and regulatory framework within the host country; (3) legal regulatory support to facilitate the import/export of nuclear technology; (4) overall leadership in the dialogue with the public regarding the need and accountability for nuclear power; and (5) stakeholder engagement across government agencies, applicable nonnuclear regulatory authorities, and industry.

Unpredictable government action, as evidenced most recently with Germany's post-Fukushima decision in 2011 immediately to permanently shut down multiple NPPs that were older and to revoke operating license extensions for the remaining

NPPs that were allowed to remain in operation, creates uncertainty for financing entities, who are trying to model NPPs and to determine the levelized cost of electricity and projected revenues over a long operating period. Thus, lenders will have to assess whether the host government has a sustained, long-term commitment to nuclear power.

The host government, through a commitment to transparency, engagement, and international practices (in the areas of safety, security, safeguards, and nuclear liability), can create the necessary alignment between the public and the nuclear power program. As noted earlier, changing public attitudes (the United States, Germany, Italy, Japan, Austria, India, the Philippines) can have dramatic impacts on individual projects (the United States, India, the Philippines, Austria) and overall nuclear power programs (Germany, Italy, Japan, Austria).

4.1.4 Host country nuclear regulatory authority

Lenders do not have the ability to monitor the NPP on a constant basis. Through their technical advisor, the lenders can make assessments of the project at various points in time, but the lenders look to the host country nuclear regulatory authority to monitor the NPP during development, construction, and operation. The nuclear regulator must be the "adult in the room" during the course of the project, and the lenders will want to have confidence in the regulator to exercise proper oversight and authority, stepping in when the safety case is put in jeopardy.

To play this oversight role, the nuclear regulatory authority must be independent within the host government's structure. Furthermore, the key technical personnel that staff the regulator must be experienced, not just as nuclear regulators, but they must also understand the technology that they are to regulate. In addition, the nuclear regulatory authority must have the tools — the authority and resources — to take regulatory action. Finally, despite the independence, the competence, and the means, the nuclear regulatory authority must clearly demonstrate the willingness to act — to take corrective measures — when the situation warrants such intervention.

This combination of factors is critical to the regulatory role. Lenders will need to review the capabilities of the host country's nuclear regulator, making a determination as to whether such regulator serves the requisite "confidence-building measure" for the NPP. Within this evaluation, the lenders will also want to assess predictability, given that regulatory uncertainty — in the form of delays and changes — has been one of the main contributing factors to projects running over budget and over schedule. Thus, while the lenders will place great importance on the regulatory function from a safety perspective, the lenders will also look to see that regulatory risk is properly allocated and mitigated within the overall project development plan. As a final note, for newcomer countries, this evaluation will be particularly challenging, and the host governments will need to take additional programmatic measures in terms of regulatory and human resources development, as well as risk allocation vis-à-vis the project, to instill confidence in the regulatory process.

In addition to the host country nuclear regulatory authority, the IAEA review process can serve to provide further oversight of the nuclear power program in the country. These reviews include:

- the Integrated Nuclear Infrastructure Review ("INIR");
- the Integrated Regulatory Review Service ("IRRS"); and
- the Operational Safety Review Team ("OSART").

Such reviews can support the goals and responsibilities of both the nuclear planning authority and the nuclear safety regulatory authority. In turn, such reviews can support the reputational risk review that is done by the lenders.

4.1.5 Sustainability

Sustainability analysis involves both environmental and social considerations. Such matters have risen in importance to the financial community in recent years, and nongovernmental organizations have used such topics to attack potential and existing NPPs. Such matters will necessarily bring project lifecycle considerations to the lenders' analysis of the project, as lenders will look to see that the NPP planning includes a spent fuel/nuclear waste plan and a decommissioning plan, demonstrating the lenders' desire to look beyond the tenor of the debt. Sustainability, in particular, is a combination of both art and science, where, oftentimes, there is no clear solution that is measurable and quantifiable; instead, the compliance plan can be qualitative, not quantitative, and, thus, much more difficult to resolve.

Sustainability considerations can be applied through a number of mechanisms, with each a function of the financial institution and/or the location of the project. As an example, the Equator Principles (http://equator-principles.com/) are a credit risk management framework for determining, assessing, and managing environmental and social risk in project finance transactions. Related policies applied to NPPs include the following:

1. International Finance Corporation's ("IFC")'s Performance Standards on Social and Environmental Sustainability and Environmental Health and Safety Guidelines
   The IFC requires its clients to apply to manage environmental and social risks and impacts, as part of the IFC's overall commitment to sustainable development.
2. "OECD's" Revised Council Recommendations on Common Approaches on the Environment and Officially Supported Export Credits
   A set of recommended common approaches for OECD Member States with respect to addressing environmental issues relating to exports of capital goods and services and the locations to which these are destined.
3. ECA-specific environmental and social guidelines
   See earlier for example of Export-Import Bank of the United States. For the example of Japan Bank for International Cooperation, see https://www.jbic.go.jp/en/business-areas/environment.html.

Each of these guidelines will be applied in varying degrees, depending on the types of lending institutions involved in the lender group, but, for planning purposes, project developers will need to evaluate each of these criteria at an early stage of the project — preferably before bid solicitation (or technology selection in a sole source scenario) — to ensure that the NPP is being developed in accordance with such requirements (if international financing is desired), with special consideration being given to where such standards exceed local law requirements.

As an additional matter, for projects being developed in UNECE Member States (largely in Europe), host governments must also comply with the Espoo and Aarhus Conventions (discussed earlier), to the extent they are Member States thereto. The Espoo and the Aarhus Conventions are implicated whenever an infrastructure project is being developed in UNECE Member States. The Espoo Convention focuses on cross-border environmental impact assessments, whereas the Aarhus Convention focuses on public participation and access to information, with emphasis on government accountability, transparency, and responsiveness, all within the environmental framework. For example, the current NPP under construction in Belarus (as well as the planned NPP in Kaliningrad that is no longer proceeding) has been challenged by Lithuania as failing to comply with the requirements of the Espoo Convention.

4.1.6 Nuclear liability and insurance

In the event of a nuclear incident at an NPP, where a radiological release occurs from containment, such release has the potential — in the case of a release on the scale of Chernobyl or Fukushima — to cause physical injury or property damage to third parties. As a precursor to this discussion, it is important to understand that an examination of nuclear liability does not involve damage to the asset itself. Such damage to the asset is covered within the contract for construction of the NPP. However, damage to third parties is outside the bounds of the contract and, due to a unique set of conditions, requires special treatment.

Under the nuclear liability conventions discussed in Chapter 8, there are several key principles that govern such conventions. For the purpose of reputational risk analysis, three are noteworthy: (1) strict and channeled liability to the licensed nuclear operator (known as "legal channeling" and distinguishable from the US Price–Anderson system of "economic channeling"); (2) equal standing of claims, regardless of nationality, domicile, or residence; and (3) mandatory financial coverage of the operator's liability.

Despite the breadth of such principles, two key concerns remain. First, no international nuclear liability convention has ever been tested in a court of law. Second, such conventions are only as effective as the membership covered thereunder (in other words, if cross-border damage is suffered in a neighboring country that is not a fellow treaty member, then legal channeling does not occur and the project participants are exposed to claims in such neighboring jurisdictions and without the benefit of the limits of liability specified under the applicable treaty; we refer to this risk as "gap risk").

For lenders, the analysis on nuclear liability is twofold. First, lenders will want to be sure that a means of protection exists for third parties, whereby a path to and an assured (and insured) source of recovery are available. Second, from an economic/risk perspective, lenders will want to be sure that the financing is structured in such a fashion to address gap risks for the lenders as project participants.

5. Part 5: concluding thoughts

This chapter has examined the 3S regime of safety, security, and safeguards. While 3S analysis is important to global nuclear governance and the national responsibilities attendant therewith, the 3S analysis also impacts reputational issues for NPP development and financing. The reality that this subjective, reputational analysis can impact whether or not a prospective NPP will go forward is instructive for prospective countries looking to embark on nuclear power programs.

For the successful financing of an NPP, technological, political (both domestic and international), reputational, and economic issues must be dealt with in a holistic manner, whereby the financing entities must be able to answer the question, "Is this a good project?" — a very subjective question that must incorporate very specific nuclear considerations and risk assessments that go beyond the more basic question of how debt payments will be serviced. Reputational risk issues encompass these holistic considerations, wherein the potential NPP is assessed both at the project and country levels to ascertain whether or not the project will align with international best practices. These nonfinancial considerations underscore the importance of larger, subjective issues in the development and financing of an NPP.

More broadly, the 3S regime underscores the need for cooperation, compliance, monitoring, and oversight across the global nuclear industry to ensure that the weaponization of nuclear technology is prevented, while promoting the beneficial aspects of nuclear technology that can promote and enhance the quality of human life. In essence, promoting industry best practices and the responsible use of nuclear technology is at the core of these international regimes. These regimes, too, are a reflection of the interdependence of the global nuclear industry — a unique industry that, from a legal perspective, combines both international treaty/convention regimes with national laws that align to create a two-tiered framework within which the 3S concept is managed.

References

Amendment to the Convention on the Physical Protection of Nuclear Material, May 9, 2016. Retrieved January 1, 2020, from: https://www.iaea.org/sites/default/files/infcirc274r1m1.pdf.

Comprehensive Nuclear-Test-Ban Treaty, 1996. Retrieved January 1, 2020, from: https://www.ctbto.org/fileadmin/user_upload/legal/CTBT_English_withCover.pdf.

Comprehensive Nuclear-Test-Ban Treaty Preparatory Commission, n.d. Status of Signature and Ratification. Retrieved January 1, 2020, from: https://www.ctbto.org/the-treaty/status-of-signature-and-ratification/.

Convention on Access to Information, June 25, 1998. Public Participation in Decision Making and Access to Justice, 1998 in Environmental Matters. Retrieved January 1, 2020, from: https://www.unece.org/fileadmin/DAM/env/pp/documents/cep43e.pdf.

Convention on Assistance in the Case of a Nuclear Accident or Radiological Emergency, November 18, 1986. Retrieved January 1, 2020, from: https://www.iaea.org/sites/default/files/infcirc336.pdf.

Convention on Early Notification of a Nuclear Accident, November 18, 1986. Retrieved January 1, 2020, from: https://www.iaea.org/sites/default/files/infcirc335.pdf.

Convention on Environmental Impact Assessment in a Transboundary Context, 1991. Retrieved January 1, 2020, from: https://www.unece.org/fileadmin/DAM/env/eia/documents/legaltexts/Espoo_Convention_authentic_ENG.pdf.

Convention on Nuclear Safety, July 5, 1994. Retrieved January 1, 2020, from: https://www.iaea.org/sites/default/files/infcirc449.pdf.

Convention on the Physical Protection of Nuclear Material., November 1979. Retrieved January 1, 2020, from: https://www.iaea.org/sites/default/files/infcirc274.pdf.

International Atomic Energy Agency, 2003. Handbook on Nuclear Law. Retrieved January 1, 2020, from: https://www-pub.iaea.org/MTCD/publications/PDF/Pub1160_web.pdf.

International Atomic Energy Agency, December 31, 2019a. Additional Protocol Status. Retrieved January 1, 2020, from: https://www.iaea.org/sites/default/files/20/01/sg-ap-status.pdf.

International Atomic Energy Agency, December 5, 2019b. Amendment to the Convention on the Physical Protection of Nuclear Material Status. Retrieved January 1, 2020, from: https://www-legacy.iaea.org/Publications/Documents/Conventions/cppnm_amend_status.pdf.

International Atomic Energy Agency, September 18, 2019c. Convention on Assistance in the Case of a Nuclear Accident or Radiological Emergency Status. Retrieved January 1, 2020, from: https://www-legacy.iaea.org/Publications/Documents/Conventions/cacnare_status.pdf.

International Atomic Energy Agency, September 18, 2019d. Convention on Early Notification of a Nuclear Accident Status. Retrieved January 1, 2020, from: https://www-legacy.iaea.org/Publications/Documents/Conventions/cenna_status.pdf.

International Atomic Energy Agency, September 18, 2019e. Convention on Nuclear Safety Status. Retrieved January 1, 2020, from: https://www-legacy.iaea.org/Publications/Documents/Conventions/nuclearsafety_status.pdf.

International Atomic Energy Agency, December 5, 2019f. Convention on the Physical Protection of Nuclear Material Status. Retrieved January 1, 2020, from: https://www-legacy.iaea.org/Publications/Documents/Conventions/cppnm_status.pdf.

International Atomic Energy Agency, September 18, 2019g. Joint Convention on the Safety of Spent Fuel Management and on the Safety of Radioactive Waste Management Status. Retrieved January 1, 2020, from: https://www-legacy.iaea.org/Publications/Documents/Conventions/jointconv_status.pdf.

International Atomic Energy Agency, 2019h. Nuclear Power Reactors in the World 2019. Retrieved January 1, 2019, from: https://www-pub.iaea.org/MTCD/Publications/PDF/RDS-2-39_web.pdf.

International Atomic Energy Agency, n.d.(a). Additional Protocol. Retrieved January 1, 2020, from: https://www.iaea.org/topics/additional-protocol.

International Atomic Energy Agency, n.d.(b). Safeguard Agreements. Retrieved January 1, 2020, from: https://www.iaea.org/topics/safeguards-agreements.

International Convention for the Suppression of Acts of Nuclear Terrorism, April 13, 2005. Retrieved January 1, 2020, from: https://treaties.un.org/doc/Treaties/2005/04/20050413%2004-02%20PM/Ch_XVIII_15p.pdf.

International Finance Corporation, 2007. Environmental, Health, and Safety General Guidelines. Retrieved January 1, 2020, from: https://www.ifc.org/wps/wcm/connect/29f5137d-6e17-4660-b1f9-02bf561935e5/Final%2B-%2BGeneral%2BEHS%2BGuidelines.pdf?MOD=AJPERES&CVID=jOWim3p.

Joint Convention on the Safety of Spent Fuel Management and on the Safety of Radioactive Waste Management, December 24, 1997. Retrieved January 1, 2020, from: https://www.iaea.org/sites/default/files/infcirc546.pdf.

Maastricht Recommendations on Promoting Effective Public Participation in Decision-Making in Environmental Matters: Prepared under the Aarhus Convention, 2015. Retrieved January 1, 2020, from: https://www.unece.org/fileadmin/DAM/env/pp/Publications/2015/1514364_E_web.pdf.

Report of the United Nations Conference on the Human Environment, June 1972. Retrieved January 1, 2020, from: http://www.un-documents.net/aconf48-14r1.pdf.

Statute of the International Atomic Energy Agency, 1989. Retrieved January 1, 2020, from: https://www.iaea.org/sites/default/files/statute.pdf.

Treaty on the Non-proliferation of Nuclear Weapons, April 22, 1970. Retrieved January 1, 2020, from: https://www.iaea.org/sites/default/files/publications/documents/infcircs/1970/infcirc140.pdf.

United Nations Economic Commission for Europe, 2013. Report of the Task Force on Public Participation in Decision-Making on its Fourth Meeting. Retrieved January 1, 2020, from: https://www.anccli.org/wp-content/uploads/2014/07/Luxembourg-roundtable-report.pdf.

United Nations Economic Commission for Europe, 2017. Good Practice Recommendations on the Application of the Convention to Nuclear Energy-Related Activities. Retrieved January 1, 2020, from: https://www.unece.org/fileadmin/DAM/env/eia/Publications/2017/1734724_ENG_web.pdf.

United Nations Economic Commission for Europe, 2019. Guide to the Aarhus Convention Compliance Committee. Retrieved January 1, 2020, from: https://www.unece.org/fileadmin/DAM/env/pp/compliance/CC_Guidance/Guide_to_the_Aarhus_Convention_Compliance_Committee__2019.pdf.

United Nations Economic Commission for Europe, n.d. Environmental Policy Background. Retrieved January 1, 2020, from: https://www.unece.org/env/pp/ccbackground.html.

United Nations Office for Disarmament Affairs, n.d. Treaty on the Non-Proliferation of Nuclear Weapons status. Retrieved January 1, 2020, from: http://disarmament.un.org/treaties/t/npt.

Civil liability in the event of a severe nuclear disaster

8

Jonathan Bellamy, C.Arb

Barrister and Chartered Arbitrator, London, United Kingdom

1. Introduction

1.1 Legal issues

This chapter addresses the topic of civil liability in the event of a severe nuclear disaster on the operator of a nuclear installation in countries with active nuclear new build programs.[1] The analysis focusses on countries presently active in nuclear new build, including China, Russia, India, the United States, the United Kingdom, and new entrants such as the United Arab Emirates (UAE).

An established, predictable legal framework, coupled with the availability of appropriate insurance, for an operator's civil liability in the event of a severe nuclear disaster is an essential part of the operation of a nuclear program for all stakeholders, including national governments, operators, supply chain contractors, and the populations concerned.

Being liability cover, the appropriate insurance will depend on the underlying legal liabilities of the operator in the jurisdiction concerned. The underlying legal regime is a function of the applicable international nuclear liability regime, if any, and the national nuclear liability regime.

The principal drivers for an international nuclear liability regime have been and remain the potential transnational effects of a severe nuclear disaster and the uncertain position in general international law in relation to legal liabilities as between states, state immunity, and jurisdiction. An illustration of this uncertainty is the fact that, following the serious nuclear incident in 1986 at Chernobyl, then situated in the USSR, which caused neighboring nation states to incur substantial cleanup costs, no claims were made against the USSR in international law. A nonnuclear power state that may be affected by a nuclear incident in a neighboring nuclear state has an obvious direct interest in a robust international nuclear liability regime to which both states are party.

This chapter identifies the volume and geography of current nuclear new build programs in the world and the key legal principles from the competing international

---

[1] It excludes analysis of civil liabilities arising from the transport of nuclear materials.

Advanced Security and Safeguarding in the Nuclear Power Industry. https://doi.org/10.1016/B978-0-12-818256-7.00008-8

nuclear liability regimes and focusses on the legal diversity of a number of relevant national regimes.

2. Nuclear new build programs

2.1 Volume

In the past decade, the number of civil nuclear reactors operable in the world has remained at a figure between 434 and 450. At the time of this writing, there are globally 441 operable reactors, 54 reactors under construction, and 109 in the planning stage.

The relative constancy in the first two figures is explicable because, during this period, the number of reactors retired from service has approximately equaled the number of reactors constructed and becoming operable. In the extended 20-year period 1996−2016, globally, 80 reactors were retired and 96 started operation. There has however been a significant recent reduction in the number of civil nuclear reactors in the planning stage to 109 in 2019, the lowest in a decade, and down more than 28% in 2018.

As Table 8.1 indicates, at any given time during the past decade, there have been between 54 and 73 nuclear reactors under construction. In the past 5 years, the figure for reactors under construction has been between 54 and 73. In the past decade, the number of reactors in the planning stage has been between 151 and 172 but has recently dropped to 109. In the past 5 years, this range has been between 109 and 172.

2.2 Geographic distribution

A significant number of countries with established operable nuclear programs do not at present have a new build program. In some such countries, e.g., Germany, it is unlikely, for reasons of public and political opinion, that any new reactors will be constructed in the foreseeable future. In others, such as the United Kingdom, there is now an active new build program, albeit currently in the planning stage.

Table 8.1 Nuclear new build: global figures: March 2020.

Year	Reactors operable	Reactors under construction	Reactors planned
2019	441	54	109
2018	450	58	153
2017	446	59	160
2016	444	62	172
2015	437	66	168
2014	434	73	172
2013	434	67	159
2012	441	60	155
2011	441	60	155
2010	439	57	151

Definitions: Operable, connected to the grid; Under construction, first concrete for the reactor poured; Planned, approvals, funding, or commitments in place, most expected to be in operation in the 2020s. World Nuclear Association: Facts & Figures March 2, 2020 (www.world-nuclear.org/information-library/facts-and-figures/world-nuclear-power-reactors-and-uranium-requireme.aspx).

The political and economic drivers for nuclear build programs are various and include increased energy demands from rising populations in developing countries, energy security, reduction of dependence on foreign suppliers of energy (e.g., natural gas), diversification of energy source, and carbon emission reduction.

As Table 8.2 demonstrates, there are presently 26 countries with active nuclear new build programs, which are defined for this purpose and in this chapter as being a country with one or more new nuclear reactors under construction or in planning (as those terms are defined earlier).

These data show that, at the present time, five countries account for over 50% of the number of new reactors under construction (in descending order of significance): China, India, Russia, Republic of Korea, and the UAE. Of these reactors, over 40% are in the most active three countries: China, Russia, and India. Neither the Republic of Korea nor the UAE have further new reactors in planning at the present time. Over 50% of the new reactors under construction are in Asia.

In addition, five countries account for over 75% of new nuclear reactors in planning (again in descending order of significance): China, Russia, India, the United States, and the United Kingdom. Of these reactors, over 70% will be in three countries: China, Russia, and India.

2.3 New entrants

The data in Table 8.3 show that at the present time about one in six new nuclear reactors under construction have been commissioned by new entrants into the civil nuclear power sector. If one excludes new reactors under construction in China, Russia, and India, the figure for new entrants would be about 30% of the global total.

The data also show that at the present time about 1 in 12 new nuclear reactors in the planning stage will be commissioned by new entrants into the civil nuclear power sector. Again, the figure for new entrants, excluding new reactors in planning in China, Russia, and India, would be over 25%.

3. The international nuclear liability regimes

3.1 Origins

A nuclear incident, and in particular a severe nuclear disaster, is likely to have a transnational effect. National laws are generally unable to deliver transnational protection for parties injured or damaged. From the early years of the development of civil nuclear operations after World War II, it was understood that nuclear countries should cooperate by way of international nuclear treaties to meet the transnational effects of the nuclear industry. This was appreciated first in western Europe under the auspices of the Organisation for Economic Co-operation and Development (OECD) and then globally under the auspices of the International Atomic Energy Agency (IAEA).

In order to encourage contractors and suppliers to participate in the development of the civil nuclear industry, it was necessary to provide appropriate limits of

Table 8.2 Nuclear new build programs: by country figures, March 2020.

Country	Reactors operable	Reactors under construction (percentage of total where >5%)	Reactors planned (percentage of total where >5%)	International liability conventions applicable
Argentina	3	1	1	VC, RVC, CSC
Armenia	1	0	0	VC
Bangladesh	0	2	0	
Belarus	0	2	0	VC
Brazil	2	1	0	VC
Canada	19	0	0	CSC
China	47	12 (22.2%)	42 (38.5%)	
Czech Republic	6	0	2	VC, JP, CSC
Egypt	0	0	2	VC, JP
Finland	4	1	1	PC, BSC, RPC, RBSC
France	57	1	0	PC, BSC, JP, RPC, RBSC
Hungary	4	0	2	VC, JP
India	22	7 (13%)	14 (12.8%)	CSC
Indonesia	0	0	1	CSC
Iran	1	1	1	
Japan	33	2	1	CSC
Korea (South)	24	4 (%)	0	
Pakistan	5	2	1	
Romania	2	0	2	VC, JP, RVC, CSC
Russia	38	4 (7.4%)	24 (22%)	VC
Slovakia	4	2	0	VC, JP
Turkey	0	1	3	PC, JP
Ukraine	15	0	2	VC, JP, CSC
The United Arab Emirates	0	4 (7.4%)	0	RVC, JP, CSC
The United Kingdom	15	1	3	PC, BSC, RPC, RBSC
The United States	96	4	3	CSC

BSC, *Brussels Supplementary Convention of 1963*; CSC, *IAEA Convention on Supplementary Compensation for Nuclear Damage*; JP, *1988 Joint Protocol Relating to the Application of the Vienna Convention and the Paris Convention*; PC, *Paris Convention on Third Party Liability in the Field of Nuclear Energy of 1960*; RBSC, *Revised Brussels Supplementary Convention of 2004*; RPC, *Revised Paris Convention: 2004 Protocol (not yet in force)*; RVC, *1997 Revised Vienna Convention on Civil Liability for Nuclear Damage*; VC, *Vienna Convention on Civil Liability for Nuclear Damage of 1963*. *World Nuclear Association: Facts & Figures March 2020 (www.world-nuclear.org/information-library/facts-and-figures/world-nuclear-power-reactors-and-uranium-requireme.aspx).*

Table 8.3 Nuclear new build programs: new entrant figures, March 2020.

Country	Reactors operable	Reactors under construction	Reactors planned	International liability conventions
Bangladesh	0	2	0	
Belarus	0	2	0	VC, RVC
Egypt	0	0	4	VC, JP
Indonesia	0	0	1	RVC, CSC
Turkey	0	1	3	PC, JP
The United Arab Emirates	0	4	0	RVC, JP, CSC
Total	0	9	8	
Global total	441	54	109	
% Total	0	16.7%	7.3%	

It excludes analysis of civil liabilities arising from the transport of nuclear materials.
CSC, IAEA Convention on Supplementary Compensation for Nuclear Damage; JP, 1988 Joint Protocol Relating to the Application of the Vienna Convention and the Paris Convention; PC, Paris Convention on Third Party Liability in the Field of Nuclear Energy of 1960; RVC, 1997 Revised Vienna Convention on Civil Liability for Nuclear Damage; VC, Vienna Convention on Civil Liability for Nuclear Damage of 1963.
World Nuclear Association: Facts & Figures March 2020.

liability. Exporting governments were also concerned that they may be liable for the loss caused by materials, equipment, and services supplied by their nationals. In the words of the IAEA, the international conventions, and the Basic Legal Principles discussed later, "reflect, on the one hand, an early recognition of the need for a stronger, more equitable system of loss distribution, in order to better protect the victims of nuclear incidents, and on the other hand, a desire to encourage the development of the nuclear industry."

3.2 Regimes

From the legal perspective, the scope and requirements of insurance cover for civil liability in the event of a severe nuclear disaster is a function of the potential liabilities in the relevant legal jurisdictions. In the nuclear liability sector, the potential civil liabilities for nuclear damage originate in both international law and national law. Insurance cover for nuclear activity in any given jurisdiction must, therefore, take into account potential liabilities from both.

There is no single international legal regime, even among countries that have developed nuclear programs.

Several of the most important civil nuclear powers, including the United States and China, rely on their own national legislation to set out civil liabilities and insurance obligations on nuclear operators.

A second group of countries, including many with established nuclear programs, are party to the Vienna Convention on Civil Liability for Nuclear Damage of 1963 ("the Vienna Convention") made under the auspices of the IAEA. In 1997, a Protocol to amend the Vienna was passed, which a small number of Vienna Convention countries have joined.

A third group of countries, also including many with established nuclear programs, are party to the Paris Convention on Third Party Liability in the Field of Nuclear Energy of 1960 ("the Paris Convention"), a treaty made under the auspices of the OECD. The Paris Convention was supplemented in 1963 by the Brussels Convention.

Parties to the Paris Convention are not parties to the Vienna Convention and vice versa. Some, but by no means all, of the parties to each of these conventions are parties to the Joint Protocol relating to the application of the Vienna Convention and the Paris Convention.

A fourth group of countries have joined the IAEA Convention on Supplementary Compensation for Nuclear Damage ("CSC"), which entered into force in April 2015 following ratification by Japan. The CSC enjoys the support of the United States. At the present time, no Paris Convention state is a party to the CSC or vice versa. The CSC is an instrument open to all states regardless of whether they are parties to any existing nuclear liability conventions or have nuclear installations on their territories. A state that is not party to either the Paris Convention or the Vienna Convention is obliged to implement national laws consistent with an Annex to the CSC. It has been said that the CSC has been developed as "an umbrella for the other international liability conventions and to provide the basis for a global nuclear liability regime that could attract broad adherence from countries with and without nuclear power plants" (IAEA, 2015a).

3.3 Key legal principles

Under both the Paris Convention and the Vienna Convention, the following legal principles ("the Seven Basic Principles") are central:

- Liability is channeled exclusively to the operator of the nuclear installation ("channelization"). The principle of channelization imposes liability on an operator to the exclusion of any other parties, such as suppliers, who would otherwise be potentially liable.
- Liability of the operator is absolute, i.e., the operator is held liable irrespective of fault, except for "acts of armed conflict, hostilities, civil war, or insurrection."
- Liability of the operator is limited in amount. Under the Vienna Convention, the upper ceiling for operator liability is not fixed but may be limited by legislation in each state. The lower limit may not be less than US $5 million. Under the 1997 Revised Vienna Convention, the lower limit was raised for members to not less than 300 million Special Drawing Rights (SDRs).[2] Under the 1960 Paris Convention, liability is limited to not more than 15 million SDRs and not less

---

[2] An SDR is the unity of currency of the International Monetary Fund, approximately equal to 1.5 US dollars.

than SDR 5 million. This figure was increased by the Brussels Supplementary Convention to a total of 300 million SDRs, including contributions by the Installation State of up to SDR 175 million and other parties to the convention collectively on the basis of their installed nuclear capacity for the balance. By the 2004 Protocol, the parties to the Paris Convention have agreed an amending protocol that will increase the upper limit for liability as follows: operators (insured), €700 million; Installation State (public funds), €500 million; and collective state contribution (Brussels), €300 million, with the total being €1500 million.

- Liability is limited in time. Generally, compensation rights are extinguished under both conventions if an action is not brought within 10 years. In addition, states may not limit the operator's liability to less than 2 years, under the 1960 Paris Convention, or 3 years, under 1960 Vienna convention, from the time when the damage is discovered.
- The operator's liability must be guaranteed by insurance or other financial security for an amount corresponding to its liability or the limit set by the Installation State. Beyond this level, the Installation State may not only provide public funds but also have recourse to the operator.
- Nondiscrimination of victims on the grounds of nationality, domicile, or residence.
- Jurisdiction over actions lies exclusively with the courts of the Contracting Party in whose territory the nuclear incident occurred.

3.4 Convention on supplementary compensation for nuclear damage

The CSC was formulated to address criticisms of the original Paris and Vienna Conventions regarding low liability limits, the lack of a supplementary compensation fund, short statutes of limitations, overly restrictive types of covered damages, and limited territorial scope.

There are presently 10 parties to the CSC. As Table 8.2 indicates, these Contracting Parties comprise Argentina, Canada, Czech Republic, India, Indonesia, Japan, Romania, Ukraine, UAE, and the United States. These parties account for almost 50% of the operable nuclear reactors and 25% of the nuclear reactors under construction in the world. It is now a global rather than a regional treaty.

By the definition of Nuclear Damage in Article 1, the CSC applies to loss of life or personal injury, loss of or damage to property, economic loss or damage arising from either of the above, costs of measures of reinstatement of impaired environment, loss of income deriving from an economic interest in any use or enjoyment of the environment incurred as a result of a significant impairment of that environment, and the costs of preventative measures and any other economic loss, other than any caused by impairment of the environment, if permitted by the general law of civil liability of the competent court.

By Articles XVIII.1 and XIX.1, the CSC provides that instruments of ratification, acceptance, approval, or accession will only be accepted from a state that is a party to either the Vienna Convention or the Paris Convention or a state that declares that its national law complies with the provisions of the Annex to the convention, provided that, in the case of a state having on its territory a nuclear installation as defined in the Convention on Nuclear Safety of June 17, 1994, it is a Contracting State to that convention. For parties who are not party to either the Paris Convention or the Vienna Convention, the Annex states that it "shall ensure that its national legislation is consistent with the provisions laid down in this Annex" (IAEA, 2015b). However, a Contracting Party having no nuclear installation on its territory is required to have only that legislation that is necessary to enable it to give effect to its obligations under the CSC.

The CSC is therefore a means by which states may align their national laws relating to civil liability for nuclear damage without necessarily joining either the Paris Convention or the Vienna Convention. A number of current members of the CSC, including Canada, India, Japan, and the United States are not members of either of these conventions. This alignment impacts the scope of nuclear liability insurance cover required in these jurisdictions.

The CSC provides two levels of compensation: the national compensation amount (Article III.1[a]) and the supplementary compensation amount (Article III.1.[b]).[3] The national compensation amount of at least 300 million SDRs available for compensation is the obligation of the Installation State. The supplementary compensation amount is not a fixed amount and is a shared obligation of the Contracting Parties whose contributions are assessed by reference to a treaty formula (Article IV), which is a function of the number of Contracting Parties and those with nuclear reactors.

Article III.2(a) provides that the first tier of compensation from the Installation State shall be "distributed equitably without discrimination on the basis of nationality, domicile or residence." Article III.2(a) provides that the second tier of compensation provided by the Contracting Parties shall be distributed in the same manner but subject to Articles V and XI.(b).

In relation to geographic territory, Article V provides that the second-tier funds apply to nuclear damage suffered in the territory of the Contracting Parties, in the maritime areas beyond the territorial sea of a Contracting Party, and in or about the exclusive economic zone or on the continental shelf of a Contracting Party.

In relation to geographic allocation, Article XI(b) states that (1) 50% of the funds shall be available to compensate claims for nuclear damage suffered in or outside the Installation State and (2) 50% of the funds shall be available to compensate claims for

---

[3] The supplementary compensation amount is to be used only if the national compensation amount is exhausted.

nuclear damage suffered outside the territory of the Installation State to the extent that such claims are uncompensated under (1). The result is that at least half of the second-tier compensation fund must be reserved for use for transnational damage.

The potential uncertainty as to the level of a Contracting Party's obligation to contribute to the supplementary compensation amount has been put forward by the Paris Convention states as a reason not to join the CSC. However, for nonnuclear states with neighboring nuclear states, the attractions of the CSC are plain. To take the example of a nonnuclear state such as Singapore situated near a future nuclear state such as Indonesia, the attractions of the extended definition of Nuclear Damage and reservation of 50% of the second-tier compensation fund for transnational loss and damage are obvious.

By Article 2 of the Annex, the CSC requires a Contracting Party's national law to

- provide for strict liability in the event of a nuclear incident where there is substantial nuclear damage off the site of the nuclear installation where the incident occurs,
- require the indemnification of any person other than the operator liable for nuclear damage to the extent that person is legally liable to provide compensation, and
- ensure the availability of at least 1000 million SDRs in respect of a civil nuclear power plant and at least 300 million SDRs in respect of other civil nuclear installations for such indemnification.

Furthermore, by Article 3 of the Annex, the CSC requires that a Contracting Party's national law shall provide, among other things, that

- where both nuclear damage and other damage have been caused by a nuclear incident, all damage shall be deemed to be nuclear damage to the extent that it is not reasonably separable from the nuclear damage;
- no liability shall attach to an operator for nuclear damage caused by a nuclear incident directly owing to armed conflict, hostilities, civil war, or insurrection;
- except insofar as the Installation State may provide to the contrary, the operator shall not be liable for nuclear damage caused by a nuclear incident caused directly by a grave natural disaster of an exceptional character[4];
- the operator shall not be liable for nuclear damage to the installation itself and any other nuclear installations and to property on the nuclear installation site, which is used or to be used in connection with the installation;
- the operator shall incur no liability for damage caused by a nuclear incident outside the provisions of national law in accordance with the CSC.

---

[4] The exception for grave natural disasters of an exceptional character has been removed from the Revised Paris and Vienna Conventions: IAEA (1997, Article 6) and OECD NEA (2004, Article 1 (J)).

Furthermore, by Article 4 of the Annex, the CSC requires that a Contracting Party's national law shall provide that

- the liability of the operator may be limited by the Installation State for any one nuclear incident to either (1) not less than 300 million SDRs or (2) not less than 150 million SDRs, provided that above that amount and up to 300 million SDRs public funds shall be made available to compensate for nuclear damage;
- an Installation State may alternatively "having regard to the nature of the nuclear installation or the nuclear substances involved and to the likely consequences of an incident originating therefrom" limit the liability of the operator to 5 million SDRs, provided that above that amount and up to 300 million SDRs public funds shall be made available to compensate for nuclear damage.

In relation to financial security, by Article 5 of the Annex, the CSC requires that the Contracting Party's national law shall provide that

- the operator shall have and maintain insurance or other financial security for its liability for nuclear damage of such type and in such amount as the Installation State shall specify,
- the Installation State shall ensure the payment of proven claims against the operator for compensation for nuclear damage by providing public funds to the extent that the limit of the insurance cover or other financial security is less than the limit of liability established pursuant to Article 4 of the CSC. Where an Installation State has not imposed a limit of liability on the operator, it may establish a limit on the amount of insurance or other financial security, provided that the limit is not less than 300 million SDRs. In this case, the Installation State shall also ensure the payment of proven claims against the operator for compensation for nuclear damage by providing public funds but only subject to the amount of the insurance or financial security to be provided pursuant to this paragraph.

In relation to prescription periods, by Article 9 of the Annex, the CSC requires the Contracting Party's national law to provide, among other things, that

- a right of compensation shall be extinguished if the action is not brought within 10 years of the date of the nuclear incident,[5]
- if, however, under the law of the Installation State, the liability of the operator is covered by insurance or other financial security or by public funds for a period longer than 10 years, the Installation State may provide that the rights of compensation against the operator shall only be extinguished after a period longer than 10 years but not longer than the period of cover by insurance or other financial security.

---

[5] The Revised Paris and Vienna Conventions extend their statute of limitations to 30 years: IAEA (1997, Article 8) and OECD NEA (2004, Article 1 (I)).

In relation to right of recourse, by Article 10 of the Annex, the CSC permits the Contracting Party's national law to provide the operator with a right of recourse only where

- this is expressly provided for by a contract in writing or
- the nuclear incident results from an act or omission by an individual done with the intent to cause damage to that individual.

3.5 Nonsovereign extralegal initiatives

The purchasers and importers of nuclear power plants for new build programs are sovereign states competent to join an international legal regime. Most countries developing nuclear new build programs import nuclear power plant technology, construction, and services from a limited number of vendors and exporters based in countries with developed nuclear programs. Acting in concert, these vendors/exporters have the opportunity and commercial leverage to influence the national legal regime of Installation States.

In 2011, under the auspices of the Carnegie Endowment for International Peace, the world major civilian nuclear power plant vendors[6] joined forces as a group entitled the Nuclear Power Plant and Reactor Exporters to issue a series of Principles of Conduct governing the sale and supply of civil nuclear infrastructure and services. The principles concerning Compensation for Nuclear Damage (NuPoC) set out a voluntary code of conduct to be followed before a group member enters a contract to supply a nuclear power plant to a purchaser. The code requires a vendor/exporter to satisfy itself that the purchaser has a national legal regime for compensation for nuclear liability complying with the Seven Basic Principles, including exclusive channelization, and membership of an international legal regime, being the Paris Convention, the Vienna Convention, or the CSC.[7]

A country developing a nuclear new build program requires a robust and internationally compliant legal regime to attract vendors and exporters of nuclear plant and services. The exclusion of an operator right of recourse is the clearest example of a commercial deal-breaker for a vendor/supplier. Concerted commercial pressure from major vendor stakeholders from countries with developed nuclear programs, supported by countries such as the United States (World Nuclear News, 2016) has an important part to play in the establishment of a global nuclear liability regime including those countries developing nuclear programs.

---

[6] Including Russia (JSC Rusatom Overseas) but excluding India and China.
[7] The Principles of Conduct state in relation to the CSC that 'Such action would enable global treaty relations crucial to assure worldwide compensation and liability protection during plant operation and transnational transport.'

3.6 The advantages of a harmonized international legal regime

In procedural matters, an international legal regime avoids uncertainties relating to, amongst other things: a state operator relying on a defense of state immunity, disputes as to legal jurisdiction and applicable law and the recognition and enforcement of judgments. In substantive matters, the key legal principles of channelization, strict liability, limits of liability, nondiscrimination, consistency in heads of recoverable damage, limited right of recourse and prescription periods operate to promote certainty and consistency as explained earlier.

From a legal and an insurance perspective, a harmonized international legal regime would achieve the above and in addition promote certainty as to potential legal liabilities and, therefore, as to risks regardless of geography. The diversity of the global nuclear new program evidences the need for harmonization. With the coming in force of the CSC binding nuclear countries such as the USA, Canada, India and Japan, and the legal principles required by the Annex of a Contracting Party's national laws, as set out earlier, this article concludes that the CSC is the most likely route toward a single global nuclear liability regime.

4. Nuclear new build countries—national legal regimes

4.1 The United States

The United States, being the pioneer of civil nuclear power, has operated under the national regime of the Price-Anderson Act of 1957. The Price-Anderson Act does not fully align with international conventions in that legal channeling is forbidden by state laws, so the act allows only economic channeling, whereby the operator is economically liable but other entities may be held legally liable. The act defines the liability of anyone liable for "public liability," which is defined as "any legal liability arising out of or resulting from a nuclear incident or precautionary evacuation."[8] This definition means that, in addition to the nuclear operator, other parties, such as suppliers, are still potentially liable for nuclear damage, but the liability is channeled to the operator because these other parties are indemnified by the nuclear operator under his/her insurance coverage. The Price-Anderson Act allows the US Nuclear Regulatory Commission (NRC) to issue regulations requiring nuclear operators to waive certain tort defenses to liability and thereby to impose strict liability. The NRC has issued regulations that apply this provision through the contractual terms of the indemnity agreements, which nuclear operators are required to conclude with the NRC.

Furthermore, under the Price-Anderson Act, unless there is an "extraordinary nuclear occurrence" (ENO), nuclear operators are subject to the ordinary standards of

---

[8] (2005) 42 USC § 2014(w). The United States Code, paragraph w.

liability.[9] In such cases, the operator will not be strictly liable and a claimant will have to prove negligence, unless the relevant state law states otherwise. After the Three Mile Island incident, which was not declared to be an ENO, many states have imposed strict liability by statute.

The way US law channels liability and its high limits of liability have prevented it from joining either the Paris Convention or the Vienna Convention. In May 2008, the United States ratified the CSC and implemented it by the Energy Independence and Security Act of 2007. A provision was included in the CSC to permit the United States to join and preserve its form of channelization, and no major changes were required to the act in order to implement the CSC.[10]

4.2 China

China is not party to any international liability convention. In January 2017, China passed a nuclear safety law by Order No.73 of the President, effective from January 1, 2018. The main rules relating to nuclear liabilities had previously been set out in two reply publications by the State Council, in 1986 and in 2007. The replies were not issued as formal administrative regulations but as replies to questions raised by government agencies.

Article 90 of the Order states that the nuclear operator "shall be bound to make compensation according to the national nuclear damage liability system" unless such damage arises from war, armed conflict, or riot. The Order does not define the scope of nuclear damage compensation, e.g., environmental damage.

The national nuclear damages liability system is thought to include the reply publications and Article 70 of the 2009 Chinese Law on Tort Liability, which provides an explicit legal basis for nuclear liability: "If a nuclear accident from a civilian nuclear installation leads to third party damage, the nuclear operator shall be liable, unless it can prove the damage is caused by war or caused by the victims on purpose."

The Order adopts the principles of strict liability and of channeling liability. It permits recourse to entities providing equipment, engineering, and services where the contract between the operator and such a third party provides for the right of recourse. The operator may exercise that right against the other person after compensating the victim.

The Order obliges the nuclear operator to make "appropriate financial guarantee arrangements" in respect of its liability by "purchasing liability insurance, participating in a mutual aid mechanism, and other means". It does not refer to minimum limits of cover.

In China, in September 2007, the liability limit was increased to a figure approaching international levels at RMB 300 million (US $47 million). Where the

---

[9] (2005) 42 USC § 2014(hh). The United States Code, paragraph hh.
[10] CSC Annex art 2.

damage exceeds this amount, state indemnity up to RMB 800 million (US $125 million) is provided. Additional indemnity may be provided for an extraordinary nuclear accident. Adoption of the CSC by China would require significant increases in the limit of liability for nuclear operators and, therefore, increase in the amount of compensation available to potential victims. Under the Order, nuclear operators are required to make appropriate financial guarantee arrangements, but the legal and financial requirements relating to insurance and financial security remain unclear.

4.3 Russia

In March 2005 the Federation of Russia ratified the Vienna Convention and incorporated it into its national law. It has not to date adopted the 1997 Revised Vienna Convention or the 1988 Joint Protocol relating to the application of the Vienna and Paris Conventions. The national law of the Russian Federation legislation states by Article 1079 that parties creating risks by, among other things, the use of nuclear power shall be strictly liable for injury and the loss caused, unless the operator proves that the injury is caused by force majeure or the deliberate act of the injured party. As a matter of general tort law, the national law of Russia applies no limit on compensation. Article 1064 states that an injured party is entitled to "full compensation" from the person who caused the damage.

In Russia, therefore, civil liability on a nuclear operator is subject to the provisions of the Vienna Convention. The principles of the Vienna Convention, including strict liability and channelization, are applicable but the extended scope of compensation under the 1997 Revised Vienna Convention is not. There remains, however, uncertainty as to the relationship between the limits of liability under that convention and the presumption of unlimited liability in national law.

4.4 India

In August 2010, India passed the Civil Liability for Nuclear Damage Act. This act brought India's national nuclear liability law into line with most of the key principles of the international convention regimes described earlier. It does, however, contain some different and specific provisions that merit reference.

First, it applies only to nuclear installations owned or controlled by the Central Government of India, either directly or indirectly (The Civil Liability For Nuclear Damage Act, 2010, Section 1 (4)). Second, it excludes operator liability where damage is caused by a nuclear incident owing to "a grave natural disaster of an exceptional character" or an act of "armed conflict, hostility, civil war, insurrection or terrorism (The Civil Liability For Nuclear Damage Act, 2010, Section 5 (1ii))." Third, by Section 6 of the 2010 Act, the limit of liability is set at 300 million SDRs or such higher amounts as the Central Government may specify by notification (The Civil Liability For Nuclear Damage Act, 2010, Section 6 (1)).

However, importantly, the position on channelization is noncompliant. Section 17(b) of the 2010 Act (entitled Operator's right of recourse) states that, although the operator shall be primarily liable for any nuclear damage, it shall, after paying the compensation set out in Section 6, have a right of recourse against its supplier to include latent and patent defects in equipment, materials, and services.[11] The operator in India will be a state party, and the reluctance to exclude a claimant's right of action, and the operator's right of recourse, against other parties such as suppliers, who will likely be foreign companies, dates back to the circumstances of the Bhopal gas disaster in 1984.

Further uncertainty has been created by the terms of Section 46 of the 2010 Act,[12] not least because, before the act was passed, civil liability for nuclear damage was subject not to prior legislation but to judicially interpreted tort law.

The Government of India has endeavored to reassure international stakeholders and supply chain contractors by stating that Section 17(b) is "not a mandatory but an enabling provision" and only applies where the operator includes a right of recourse in its contract with the supplier. This, however, is contrary to the natural reading of the section. It has also stated that Section 46 applies only to operators and not to other parties. Again, this is not evident from the language of the section.

Although India ratified the CSC in February 2016, it remains unclear how it relates to the national law, in particular on the fundamental principle of channelization. A respected legal commentator on Indian nuclear law has described the 2010 Act as "flawed legislation."

4.5 The United Kingdom

The United Kingdom is a signatory to the Paris Convention of 1960 and to later amendments, including the 2004 Protocol that is not yet in force. The national law is contained in the Nuclear Installations Act 1965 (as amended) (NIA 1965). The most recent and wide-ranging amendments will bring into force in English law the provisions of the 2004 Protocol to the Paris Convention. The UK national legislation, therefore, fully reflects and implements the key legal principles of the Paris Convention, including strict liability of the operator, channelization of liability to the operator, compulsory financial security, limitation of liability in time, limitation of liability in amount, jurisdiction and applicable law, and nondiscrimination.

---

[11] The Civil Liability For Nuclear Damage Act, 2010, Chapter IV Claims and Awards, Section 17 (b): 'The operator of the nuclear installation, after paying the compensation for nuclear damage shall have a right of recourse where (a) such right is expressly provided for in a contract in writing; (b) the nuclear incident has resulted as a consequence of an act of supplier or his employee, which includes supply of equipment or material with patent or latent defects of sub-standard services; (c) the nu- clear incident has resulted from the act of commission or omission of an individual done with the intent to cause nuclear damage'.

[12] The Civil Liability For Nuclear Damage Act, 2010, Chapter VII, Section 46 Under the title 'Act to be in addition to any other law', s 46 states: 'The provisions of this Act shall be in addition to, and not in derogation of, any other law for the time being in force, and nothing contained herein shall exempt the operator from any proceeding which might, apart from this Act, be instituted against such operator'.

Of note are the new provisions in the draft amended NIA 1965 to implement the 2004 Protocol to the Paris Convention, extending the period for liability in time from 10 years to up to 30 years for personal injury, in particular cancers, caused by a nuclear incident.[13] This limitation period overrides general limitation periods for latent personal injuries in English law, which depends for their duration on the date and extent of the claimant's knowledge.

In terms of future legal development, one aspect of BREXIT will be that the United Kingdom is able to negotiate new treaty obligations unilaterally and without EU authorization. This is likely to increase the prospect of the United Kingdom joining the CSC.

The UK nuclear new build and decommissioning program, involving parties from CSC states (e.g., the United States, Canada, Japan) with whom the UK enjoys established relationships in nuclear, together with the UK international nuclear transport activities, provides an economic driver for joinder to the CSC. Joining the CSC would obviate, or at least significantly reduce, the need for UK state-backed indemnities to the United States and other Contracting Parties in such projects. Joinder would also reduce forum shopping. In the absence of CSC treaty relations between the countries, a nuclear incident in the United Kingdom involving US personnel would, for example, engage (or have the potential to engage) the jurisdiction of the US courts; see by analogy, *Cooper v TEPCO*, US Court of Appeals for the Ninth Circuit June 22, 2017 (US class action brought by US navy personnel arising from the Fukushima incident).

4.6 The United Arab Emirates

The UAE is an imminent new entrant to the group of nuclear power countries with four new reactors under construction at the Barakah nuclear power plant. The operating license for Unit 1 was issued in February 2020. In contrast to a number of other prospective new entrants, including Turkey and Bangladesh with reactors under construction, the UAE is not only a member of an international legal regime (the 1997 Revised Vienna Convention ratified in 2012 by Federal Decree No.32 of 2012) but also has national legislation in place before any of the new reactors become operable.

The UAE nuclear liability law, Federal Law by Decree No.4 of 2012, was drafted in consultation with the IAEA and was reviewed by the IAEA's legal team to ensure consistency with its guidance and the UAE international obligations under the 1997 Revised Vienna Convention. The Articles of the Federal Law expressly incorporate the relevant articles of the 1997 Revised Vienna Convention. Exclusive jurisdiction over liability claims is given to the Federal Courts of the Emirate of Abu Dhabi (UAE Federal Law, Article 12(2)), and it remains to be seen how the national law will be interpreted by the national court. By this national law, the UAE has not

---

[13] This future extended liability period is problematic for the insurance market.

only implemented the key legal principles of the international legal regime but also set the operator's limit of liability at a figure of 450 million SDRs, being 50% greater than the required minimum in the 1997 Revised Vienna Convention (UAE Federal Law, Article 5(1)). The national law requires the operator to provide insurance or other financial security in that amount and provides that, if the operator is unable to obtain the required insurance cover, either nationally or internationally, the risks covered by the insurance shall be covered directly by the state up to the required limit of liability (UAE Federal Law, Article 8(2)).

4.7 The advantages of international convergence

Many of the countries with significant nuclear new build programs have different national legal regimes governing civil liability for nuclear damage. This may be because (1) for historical reasons the countries are not party to an international legal regime (China), (2) until recently the countries were not party to an international legal regime (the United States, India), or (3) the countries are, again for historical reasons, party to different international regimes (Russia, the United Kingdom).

Countries planning and constructing a new nuclear industry are well advised to adopt an established international legal liability regime. This operates to protect all stakeholders and populations. It establishes clear legal and commercial rules on risk allocation. A positive example of this approach is the new entrant, UAE. The UAE has demonstrated its commitment to incorporate the key legal principles established internationally into its national law before completing its new build program.

The most likely route by which inconsistencies created by disparate national law regimes may be harmonized is through joinder of CSC and by national laws complying with the legal requirements in the Annex. The CSC is open to adoption to all countries, whether or not they have a nuclear industry. Given its prominence in global nuclear new build, the joinder of China to the CSC and compliance of its national law would be the most significant single step in that direction.

---

5. Conclusion

From a legal perspective, the various international regimes have resulted in a number of generally, but not universally, accepted legal principles. The scope for further harmonization has been increased by the accession of Japan and India to, and, therefore, the coming into force of, the IAEA CSC. However, there remain important differences between national legal regimes.

From an insurance perspective, certainty of the potential legal liabilities is an essential part of the underwriting process and further harmonization, whether based on the development of the CSC or otherwise, is to be welcomed.

From a political perspective, it should be remembered that no international or national legal regime has or is intended to have the capacity to provide full financial compensation following a nuclear catastrophe of the order of Fukushima. The insurer of last resort will remain to be the Installation State—a point that will be acceptable in a developed nuclear power country such as Japan but will result in a compensation vacuum in a developing nuclear power country.

References

IAEA, The International Expert Group on Nuclear Liability (INLEX): Civil Liability for Nuclear Damage: Advantages and Disadvantages of Joining the International Nuclear Liability Regime. International Atomic Energy Agency, 15 April 2015a. Available from: https://www.iaea.org/sites/default/files/17/11/liability-regime.pdf.

IAEA, Convention on Supplementary Compensation for Nuclear Damage, Preamble to Annex, International Atomic Energy Agency 15 April 2015b.

IAEA, 1997 Protocol to Amend the Vienna Convention, Article 6, International Atomic Energy Agency, 12 September 1997.

IAEA, 1997 Protocol to Amend the Vienna Convention, Article 8, International Atomic Energy Agency, 12 September 1997.

NuPoC, Nuclear Power Plant and Reactor Exporters' Principles of Conduct, entry into force 8 May 2016. Available from: http://nuclearprinciples.org/principle/compensation-fornuclear-damage (Accessed 31, March 2020)."

OECD NEA, 2004 Protocol to Amend the Paris Convention, Article 1 (J), OECD Nuclear Energy Agency, 12 February 2004.

OECD NEA, 2004 Protocol to Amend the Paris Convention, Article 1 (I), OECD Nuclear Energy Agency, 12 February 2004.

The United States Code, 2010 Edition. Title 42 - Thee Public health and Welfare. Chapter 23 — Development and Control of Atomic Energy, Division A —Atomic Energy, § 2014 Definitions. Available from: https://www.govinfo.gov/content/pkg/USCODE-2010-title42/html/USCODE-2010-title42-chap23-divsnA.htm.

The Civil Liability For Nuclear Damage Act, 2010, Chapter I Preliminary, Section 1 (4). As the nuclear industry in India is presently configured, this provision creates no difference in practice.

The Civil Liability For Nuclear Damage Act, 2010, Chapter II Liability for Nuclear Damage, Section 5 (1ii).

The Civil Liability For Nuclear Damage Act, 2010, Chapter II Liability for Nuclear Damage, Section 6 (1).

UAE Federal Law by Decree concerning. Civil Liability for Nuclear Damage, adopted 13 August 2012, https://www.ilo.org/dyn/natlex/docs/ELECTRONIC/92437/107663/F-976812755/Federal-Law-by-Decree-No-4-of-2012-Concerning-Ci.pdf.

UAE Federal Law by Decree concerning. Civil Liability for Nuclear Damage, Article 12(2), adopted 13 August 2012. Available from: https://www.ilo.org/dyn/natlex/docs/ELECTRONIC/92437/107663/F-976812755/Federal-Law-by-Decree-No-4-of-2012-Concerning-Ci.pdf.

UAE Federal Law by Decree concerning. Civil Liability for Nuclear Damage, Article 5(1), adopted 13 August 2012. Available from: https://www.ilo.org/dyn/natlex/docs/ELECTRONIC/92437/107663/F-976812755/Federal-Law-by-Decree-No-4-of-2012-Concerning-Ci.pdf.

UAE Federal Law by Decree concerning. Civil Liability for Nuclear Damage, Article 8(2), adopted 13 August 2012. Available from: https://www.ilo.org/dyn/natlex/docs/ELECTRONIC/92437/107663/F-976812755/Federal-Law-by-Decree-No-4-of-2012-Concerning-Ci.pdf.

World Nuclear News: The US Energy Secretary greeted India's signature of the CSC saying the US was 'eager to work with India, and all CSC member countries, to facilitate the use of advanced nuclear technologies developed in the United States', 5 February 2016.

Future challenges in safety, security, and safeguards

9

Tatsujiro Suzuki, MS

Professor, Doctor, Research Center for Nuclear Weapons Abolition, Nagasaki University,
Nagasaki, Japan

1. Introduction

The global nuclear industry is now facing big challenges in three major fields, i.e., safety, security, and safeguards, which are often referred as "three S" in the nuclear industry. Technical safety is not a major subject of this chapter, but security and safety have much common features to be shared. Besides, loss of public trust after the Fukushima nuclear accident in 2011 has affected all aspects of nuclear industry, including safety and security. This chapter addresses three major new challenges for both of global nuclear industry in the area of security and safeguards and explore possible policy options to overcome such challenges.

Three challenges for nuclear security are increased stockpile of fissile materials, emerging threats coming from new technologies such as cyber, and institutional challenges to internal threats/sabotages and illicit trade.

Three challenges for safeguards are increasing number of new comers to introduce nuclear power programs, dealing with noncompliance such as the ones in DPRK, Iraq, and Syria and new type of safeguards activities such as the one under the Joint Comprehensive Plan of Action (JCPOA) in Iran or possible verification schemes for nuclear disarmament.

This chapter argues that in each field, much close international cooperation and possible new norm will be necessary to confront those challenges.

2. Overview of global nuclear industry

What is the current status of global nuclear industry? According to the International Atomic Energy Agency (IAEA)'s data base, As of February 2020, total of 442 nuclear reactors (390.5GWe) are operating and 53 reactors (56.3 GWe) are under construction (IAEA, 2020). It suggests that global nuclear industry is still growing. But closer look at the market may suggest the different view.

According to the Global Nuclear Industry Status Report (Schneider et al., 2018), global nuclear industry is facing economic, environmental, and safety concerns, and

Advanced Security and Safeguarding in the Nuclear Power Industry. https://doi.org/10.1016/B978-0-12-818256-7.00009-X

its future is very uncertain. The share of nuclear power in total electricity production has declined from the peak of 17.5% in 1996 to 10.2% in 2018, and there is a declining trend in nuclear share in large nuclear power countries, such as France (78.5% in 2005 to 71.7% in 2018), Germany (30.8% in 1997 to 11.7% in 2018), Japan (25% in 2010 to 3.6% in 2017), and the United States (22.5% in 1995 to 19.3% in 2018). The impact of Fukushima nuclear accident is strong, in particular, in those liberalized markets. On the other hand, China is now a leading country in global nuclear market. China ranks first in terms of the scale of nuclear power programs under construction and is also expanding international cooperation programs under the "One Belt One Road" initiative. Russia is following China in expanding share of global nuclear market.

In short, here is the summary of global nuclear industry:

- The global nuclear reactor market is expanding, but the importance of nuclear power in total production has been declining and possibly will decline in the future especially in most advanced nuclear power countries (mostly in OECD countries).
- On the other hand, China and followed by Russia are the growing world leaders in global nuclear industry. Leadership of global nuclear industry is now shifting from traditional advanced nuclear countries to China and Russia.

What are the implications of these trends to security and safeguards of nuclear industry? This question must be addressed seriously, and the following sections will try to answer those questions.

3. Loss of public trust in nuclear safety

Public opinion on nuclear power in Japan has changed dramatically since the Fukushima accident. Perhaps the most obvious is lack of trust, especially of the government, but also of nuclear industry and nuclear experts. It underpins the challenges facing Japan's nuclear industry since the Fukushima disaster in 2011, and lack of trust is a fundamental problem.

The public has lost faith in nuclear safety. Even the establishment of a new independent Nuclear Regulation Authority in 2012 and new tougher regulatory standards have failed to recover public faith. According to the public polling done by Japan Atomic Energy Relations Organization (JAERO, 2019), 1.7% of the public trust the government and only 3.0% of the citizens trust nuclear industry. As a result, the ratio of the public who think that nuclear power should be increased or even maintained dropped to 10.1% in 2014 and further to 9.6% in 2018. Understandably, the ratio of the public who are in favor of phasing out or abolishing nuclear power immediately increased to 60.5% by 2018. The ratio of the public who believes that safety of nuclear power plant is assured is only 6.3% in 2018.

JAERO is a pronuclear organization, and thus, the changes they reported are particularly striking.

This loss of trust is the most serious challenge that nuclear policymakers and the nuclear industry now face in Japan. Despite serious efforts by the nuclear industry and NRA, the public is not assured of nuclear safety in Japan.

4. Challenges for nuclear security

Nuclear security has attracted world attention since the US President Obama took initiatives to host a series of "Nuclear Security Summit" from 2010 to 2016.[1] The Summit meetings had positive impact on nuclear security, attracting attention of world leaders to the risk of nuclear terrorism. But on the other hand, since the end of the Summit, it is also true that many things have not been achieved. Ms. Laura Holgate, who has been working in this area from the Clinton Administration and again in Obama Administration, stated that the achievements of the Security Summits include the following: (1) over 935 so-called "house gifts" that countries made commitments to improve nuclear security, especially reducing the number of countries who have fissile materials stockpile, (2) increase in International Physical Protection Advisory Service (IPPAS) missions, (3) increase in ratification of the 2005 amendment to the Convention on Physical Protection of Nuclear Materials (CPP), which increased from 20 in 2009 to 100 in 2016(Suzuki, 2018). On the other hand, Ms. Holgate also stressed that there are missions accomplished, saying that "complacency" is the biggest problem. In particular, those who did not participate in the Summit, including Russia, may not consider nuclear security as a top priority in their security policy.

It is true that risks of nuclear terrorism may not be equally felt by country leaders. But once it happened, the impact of nuclear terrorism may be worldwide. And thus, the global nuclear industry and world leaders must continue to improve its nuclear security measures. Especially, this chapter deals with the following three challenges:

1. Increased stockpile of fissile materials

Global stockpile of fissile materials (weapons usable) is still increasing. At the end of 2017, there were about 1340 tons of highly enriched uranium (HEU) and 523 tons of separated plutonium worldwide (RECNA, 2019). If you translate these quantities into the number of atomic bombs equivalent, those stockpiles are about 108,104 bombs equivalent (64 kg/bomb for HEU, Hiroshima type bomb, and 6 kg/bomb for plutonium, Nagasaki type bomb). While the stockpile of HEU is decreasing, the stockpile of plutonium is increasing primarily due to increase in civilian plutonium (291 tons now) coming from reprocessing programs of civilian nuclear fuel. Nonmilitary plutonium stockpile, including "excess plutonium from

---

[1] The first Summit took place in Washington, DC in 2010, the second on in Seoul, South Korea in 2012, the third in Hague, Netherland in 2014, and the final one in Washington, DC in 2016.

Table 9.1 Global stockpile of separated plutonium worldwide (as the end of 2017).

Stockpile of separated plutonium		
Country	Military (ton)	Non-military (ton)
Russia	94.0	93.0
United States	38.4	49.4
France	6.0	65.4
China	2.9	0.04
United Kingdom	3.2	113.0
Israel	0.92	0.0
Pakistan	0.31	0.0
India	7.07	0.4
DPRK	0.04	0.0
Japan	0	47.3
Germany	0	0.0
Other NNWS		0.0
Subtotal	152.8	368.5
Total	521.4	

Source: RECNA, December 2019. A Guide to the World's Fissile Material Inventory. https://www.recna.nagasaki-u.ac.jp/recna/bd/files/FM2019_EB.pdf. (accessed 5 March 2020).

military use," now reached to 370.1 tons, which is about 70% of total plutonium stockpile (Table-9.1).

It should be noted that the number of countries who possess fissile materials decreased from 57 countries to 24 countries (Bunn et al., 2019). The countries that eliminated fissile materials in its soil are Iraq (1992), Colombia (1996), Spain (1997), Denmark (1998), Thailand (1999), Slovenia (1999), Brazil (1999), Philippines (1999), Greece (2005), South Korea (2007), Latvia (2008), Bulgaria (2008), Portugal (2008), Libya (2009), Romania (2009), Taiwan (2009), Chile (2010), Serbia (2010), Mexico (2012), Ukraine (2012), Sweden (2012), Austria (2012), Czech Republic (2013), Vietnam (2013), Hungary (2015), Jamaica (2015), Uzbekistan (2015), Georgia (2015), Argentina (2016), Indonesia (2016), Poland (2016), Ghana (2017), and Nigeria (2018). As of the end of 2018, nine countries (China, France, India, Japan, Kazakhstan, Pakistan, Russia, the United Kingdom, and the United States) have more than 2 tons of fissile materials including those for military purposes (Japan and Kazakhstan do not have military materials). 10 countries (Belarus, Belgium, Canada, Germany, Israel, Italy, Netherland, North Korea, South Africa, and Turkey) have less than 2 tons but greater than 10 kg (only Israel and North Korea have military materials). And five countries (Australia, Iran, Norway, Switzerland, and Syria) have less than 10 kg of fissile materials (all civilian materials).

But the most pressing issue for civilian nuclear industry is increasing stockpile of separated plutonium for civilian purposes. There are only four countries (Russia,

France, the United Kingdom, and Japan) that have large commercial reprocessing program (while the United Kingdom will phase out commercial reprocessing program soon). Those four countries have total of about 280 tons, more than half of total plutonium stockpile worldwide including military ones. It is important that those plutonium civilian stockpile must be reduced or at least consolidated to minimize the risk of nuclear terrorism. In addition, stockpile of excess plutonium from nuclear disarmament (about 49 tons in the United States and 34 tons in Russia) has not been reduced. It is essential for those countries to cooperate each other to find a way to reduce the stockpile of separated plutonium.

2. Dealing with new technologies such as cyber

Although physical security against nuclear terrorism has been improving quite significantly, new threats, such as cyberattacks, are now emerging, and measures against such "soft" threats can be more challenging.

Cyberattacks could be used to facilitate the theft of nuclear materials or/and an act of sabotage that could result in significant release of radioactive materials. Cyberattacks against nuclear facilities could lead to malfunction of nuclear safety systems, which may result in severe nuclear accidents.

Nuclear operators are well aware of those risks and typically take physical measures, such as creating "air gap" (isolate nuclear control systems from outside network). The study done by the Chatham House concludes that "air gap" is not enough against cyberattacks (Livingstone et al., 2015). The summary of their conclusions are as follows:

- The conventional belief that all nuclear facilities are "air-gapped" is a myth. The commercial benefits of the Internet connectivity mean that a number of nuclear facilities now have VPN (virtual private network) connections installed, which facility operators are sometimes unaware of.
- Search engines can readily identify critical infrastructure components with such connections. Even where facilities are "air-gapped", this safeguard can be breached with nothing more than a flash drive.
- Supply chain vulnerabilities can mean that equipment used at a nuclear facility risks compromise at any stage.
- A lack of training, combined with communication breakdowns between engineers and security personnel, means that nuclear plant personnel often lack an understanding of key cybersecurity procedures.
- Reactive rather than proactive approaches to cybersecurity contribute to the possibility that a nuclear facility might not know of a cyberattack until it is already substantially underway.

According to the Nuclear Threat Initiative (NTI), there have been at least more than 20 cyber-related incidents at civilian nuclear facilities (NTI, 2020). NTI also published a report that recommends the following actions to improve cybersecurity of civilian nuclear facilities (Van Dyne, Assante, and Stoutland, 2016):

- Institutionalize cybersecurity: Implementation of robust processes and practices is essential.
- Mount an active defense: Nuclear facilities need to develop the means to respond to threats once a compromise occurs, as firewalls and airgaps have proven to be limited in their efficacy.
- Reduce complexity: Complexity is the enemy of security. The security impact of the current complex system and how they interact are not always fully understood.
- Pursue transformation: The global community is in the early stages of understanding the magnitude of the cyberthreat. There is a fundamental need for transformative research to develop hard-to-hack systems for critical applications.

Chatham House also published a report on cyberinsurance for nuclear industry, in which they suggest that all civil nuclear facilities should consider the establishment of computer security incident response (CSIR) teams, saying that the existence of a CSIR team will be a prerequisite for any facility seeking to obtain civil nuclear cyberinsurance (Leverett, 2019).

There are other new technologies that need to be paid attention. For example, drones (unmanned aerial vehicles: UAVs) can be a new threat to nuclear security, although same technologies also can *improve* nuclear security (Baylon, 2014). It is important for nuclear operators to be aware of such emerging risks as development of technologies could outpace improvement of security measures taken by the nuclear industry.

3. Institutional/organizational challenges to deal with internal threats/sabotage

One of the lessons learned from the Fukushima nuclear accident on nuclear security issues is that sabotage could lead to catastrophic radiological release. Especially spent fuel storage pool can be most vulnerable place in nuclear security. Even if nuclear operators do not have any fissile materials (HEU or separated plutonium), inadequate protection of spent fuel storage pool could have similar consequences of serious nuclear terrorism.

Internal threats and/or sabotage is difficult to prevent. In fact, nearly every publicly documented case of nuclear theft and sabotage at a nuclear facility were carried out by insiders or with the help of insiders (Bunn and Sagan, 2017).

IAEA's new guideline, INFCIRC/225/Rev. 5 noted that "insiders could take advantage of their access rights, complemented by their authority and knowledge, to bypass dedicated physical protection elements or other provisions, such as safety procedures" (IAEA, 2011).

At the 2016 Nuclear Security Summit, more than 24 countries sign on the document, "Mitigating Insider Threats," which now became formal document at IAEA, INFCIRC/908(IAEA, 2017). INFCIRC/908 lists a wide range of steps that states could take to address insider threats, but no commitment to action was clarified. Matthew Bunn et al. argue that a comprehensive, multilayered approach, rather than only one or two protections, is required to have a truly effective security system against sophisticated individuals, multiple insiders working together, and insiders working with outsiders (Bunn and Sagan, 2017). They recommend the effective strategies should include (1) background checks before granting access and ongoing monitoring after access is permitted, (2) strong incentives for staff to report any concerning behavior, or any potential vulnerabilities they observe, (3) effective programs to address employee disgruntlement, (4) regular training programs focused on protecting against insider threats, (5) nuclear material accounting that is accurate and timely enough to detect either a rapid or a protracted theft, identify when and where it happened, and establish who had access then, (6) constant surveillance of nuclear material and of vital areas that might be sabotaged, (7) physical protection systems consciously designed to handle both insider and outsider threats and inspections to ensure the effectiveness of the insider protection in place.

In short, protection measures against insider threats need to be enhanced, and attention of top management of nuclear industry is essential to keep these measures truly effective.

5. Challenges for nuclear safeguards

There are three new challenges for nuclear safeguards that the global nuclear industry is facing.

1. Growing new comers and resource constraints

The first challenge is the increased number of new comers and expanding global nuclear capacity, which require more workload and budget for IAEA. As a result, the number of nuclear facilities and the use of nuclear material continue to grow. In 2015, the IAEA safeguarded 1286 nuclear facilities and locations outside facilities, such as universities and industrial sites. IAEA inspectors carried out 2118 inspections in the field. Other statistics on IAEA include 967 nuclear materials and environmental samples collected, 883 workforce at IAEA safeguards department, 1416 surveillance cameras in operation, network of 20 qualified laboratories, and 407 satellite images (IAEA, 2016).

In the past 5 years (2010–15), 7 new safeguards agreements and 23 new additional protocols (APs) entered into force. The quantities of nuclear material under safeguard have increased by 17% and the number of nuclear facilities under safeguard by 5%. This trend will likely to continue as the number of new nuclear power programs continues to grow.

But with condition of no budget increase, safeguards department must find ways to meet growing demand with constant resources. One way is to make full use of advanced technologies, and the second is more efficient operation at the IAEA(Varjoranta, 2016).

One such project is MOSAIC (the Modernization of Safeguards Information Technology) program. The program (2015–18) is developing applications and software to optimize and modernize the safeguards operations.

2. Dealing with noncompliance

Second challenge is how to deal with "noncompliance" of safeguards agreement. Table 9.2 summarizes the past IAEA noncompliance reporting (Findley, 2015). Iraqi case in 1991 was the first case, and the number of noncompliance cases increased since 2002. The process with which IAEA reports noncompliance itself is complicated and bureaucratic, as there are different kinds of safeguards and procedures. It was often criticized as not to be objective or biased while IAEA report itself is by nature technical in detail. Sovereignty is often a major barrier to verify the important information. Based on the analysis of past cases, Findlay made the following key recommendations (Findley, 2015):

- The safeguards noncompliance process cannot be rendered machine-like and automatic. The IAEA must always have room to offer diplomatic, technical, and other creative solutions that may not have been foreseen by the Agency's founders and continuing practitioners.
- The Secretariat must preserve it apolitical, neutral, and technical reputation. It reinforces its principle the Board of Governors that must take responsibility for declaring a state to be in noncompliance.
- Despite the technical detail involved, the noncompliance process is an intensely human one involving interactions between international civil servants and representatives of governments. This fact reinforces the vital importance of well-recruited and professionally trained staff, a diplomatically skilled director general, and well-informed governors.
- The IAEA should be more transparent about all states' compliance with their safeguards obligations. A good start would be to publicly release the annual Safeguards Implementation Report (SIR).
- Member states should ensure that the Agency has cutting-edge verification technology and analytical tools, first-rate professional and support staff, and generous funding to enable it to handle increasingly complicated noncompliance case.
- Leadership is key. Choosing a politically and diplomatically astute director general, who is willing and able to take full advantage of the experience, technical skills, and advice of the Secretariat, while balancing the various interests of member states and stakeholders, is vital.

3. New type of safeguards activities such as JCPOA

The final challenge is new type of safeguards requirements from special agreements among governments and institutions.

The best example is JCPOA, which is the agreement between Iran and P5+1 (China, France, Russia, the United Kingdom, the United States, and Germany) reached on July 14, 2015. The JCPOA has various new features on verification on Iran's nuclear program, which are not typically included in the comprehensive safeguards required under the Non-Proliferation Treaty (NPT). Here are some of the key new features of verification measures under the JCPOA (The White House, 2015).

Table 9.2 IAEA noncompliance reporting.

Noncompliance case (reporting period)	Date of first report	No. of written reports[a]	Recipient of reports (no.)	No. and % publicly available
Iraq (1991—2005)	July 1991[b]	65[c]	United Nations Security Council (UNSC) (63)[d] BOG (2)[e]	65 (100%)
Romania (1992)	June 1992	0	Oral report to Board of Governors (BOG)	0
DPRK (1992 —present)	June 1992[f]	57[g]	BOG (35) GC (14) BOG and GC (8)	23[h] (40%)
Iran (2002 —present)	September 2002	53	BOG (30) BOG & UNSC (23)	53[i] (100%)
Libya (2003—04)	February 2004[j]	5	BOG (4)	4[k] (80%)
ROK (2004)	August 2004[L]	1	BOG (1)	1[m] (100%)
Egypt (2004—05)	September 2004	1	BOG (1)	1[n](100%)
Syria (2008 —present)	November 2008[o]	15	BOG (15) corrigendum (1)	15[p] (100%)

[a] *Numbers are approximate due to public inaccessibility of entire corpus of reports. Excludes oral reports.*
[b] *"Statements and Reports," INVO, http://www.iaea.org/OurWork/SV/Invo/statements.html (accessed September 30, 2015); excludes letters and other correspondence.*
[c] *27 inspection reports, nine on "destruction, removal and rendering harmless," nine on Ongoing Monitoring and Verification (OMV) and 18 consolidated reports (after April 1996); see Gudrun Harrer, Dismantling the Iraqi Nuclear Program: The Inspections of the International Atomic Energy Agency, 1991—98 (London and New York: Routledge, 2014), pp. 90—91.*
[d] *After September 1991 also sent to General Conference (GC) and BOG.*
[e] *Presented to BOG in July and September 1991.*
[f] *Inspections began May 1992 and Director General (DG) reported results orally to Board in June. First publically available report from DG to GC dated August 16, 1993.*
[g] *Based on assumption that IAEA website list is comprehensive and broken links are reports not publically available. Figure includes DG reports to the BOG publicly unavailable, but referenced in GC reports from 1993 to 2001, and reports known to exist from 2002—present, plus a presumed 10 earlier inspection reports between 1992 and 1999 http://cns.miis.edu/archive/country_north_korea/nuc/iaea9799.htm (accessed September 30, 2015).*
[h] *See "IAEA Reports," https://www.iaea.org/newscenter/focus/dprk/iaea-reports (accessed September 23, 2015). Links to some reports are broken. A link to the August 14, 2006 report from DG to GC does not appear on IAEA Reports page but does appear on IAEA and DPRK Chronology of Events page. July 3, 2007 Board Report available at http://isis-online.Org/iaea-reports/category/korean-peninsula/#2007 (accessed September 30, 2015). See also https://www.iaea.org/About/Policy/GC/GC59/Documents/(accessed September 30, 2015).*
[i] *Reports available at https://www.iaea.org/newscenter/focus/iran/iaea-and-iran-iaea-reports (accessed September 30, 2015).*
[j] *First Board report available dated February 20, 2004; report of December 22, 2003 (GOV/2003/82) not available on IAEA website. From 2005—September 2008 updates on Libya included in SIR.*
[k] *Four reports available at http://isis-online.org/iaea-reports/category/libya/(accessed September 23, 2015).*

Continued

Table 9.2 IAEA noncompliance reporting.—*cont'd*

Noncompliance case (reporting period)	Date of first report	No. of written reports[a]	Recipient of reports (no.)	No. and % publicly available

l It appears that inspections from late August 2004–November 2004 generated only one report. DG's September report to the BOG appears to have been oral.
m See http://www.globalsecurity.org/wmd/library/report/2004/index.html (accessed September 23, 2015).
n Report available at http://www.globalsecurity.org/wmd/library/report/2005/egypt_iaea_gov-2005-9_14feb2005.pdf (accessed September 23, 2015). No other reports until 2008 SIR (not publicly available).
o Total does not include oral statement made by DG to BOG in September 2008.
p Reports available (except August 28, 2013 BOG report) at https://www.iaea.org/newscenter/focus/syria/iaea-and-syria-iaea-reports (accessed September 23, 2015); August 28, 2013 Report available at http://isis-online.org/iaea-reports/category/syria/(accessed September 23, 2015).
Source: Findlay, T., October 2015. Proliferation Alert! : The IAEA and Non-Compliance Reporting, Project on Managing the Atom. https://www.belfercenter.org/sites/default/files/legacy/files/Proliferation%20Alert.pdf.

- Continuous access to nuclear activities: Not only with the IAEA have the right to a constant physical or technical presence in Iran's nuclear sites, but also it will be able to conduct regular monitoring of Iran's uranium mines, mills and its centrifuge production, assembly, and storage facilities. This means that IAEA will have surveillance of the entire fuel cycle and supply chain for Iran's nuclear program.
- Timely access to suspicious locations guaranteed: If Iran were to deny an IAEA request for access to suspicious undeclared location, a special provision in the JCPOA would trigger an access dispute resolution mechanism. If Iran and IAEA cannot resolve the access dispute within 14 days, the issue is brought to the Joint Commission, which then has 7 days to find a resolution. If Iran still will not provide access but five members of the Joint Commission determine necessary, Iran must the provide access within 3 days.
- Verification of past possible military program: On May 29, 2015, the Director General stated that "The Agency remains concerned about the possible existence in Iran of undisclosed nuclear related activities involving military related organizations"(IAEA, 2015). IAEA must report the findings to the Joint Commission.

All these new requirements make the verification (safeguards) under the JCPOA unprecedented. But it could be a good reference for future new type of verification measures such as the one required for denuclearization of DPRK's nuclear program.

6. "Safeguarding" public from nuclear risks

The title of this book says "safeguarding" rather than "safeguards." It is interesting to note the tone and implication of the two different words. As noted earlier, the biggest impact of the Fukushima nuclear accident was a loss of public trust. The accident has made public feel that they are "safeguarded" by the government and nuclear industry. This is clearly shown by the polling data noted earlier.

There are two interesting Japanese words that sound similar but are often used to emphasize the difference among them. One is "*Anzen*" that can be literally translated to "safety" in English. The other word is "*Anshin*" that is more difficult to translate but can be interpreted as "assurance" in English. "*Anzen*" (safety) can be defined as a state in that risk is within acceptable range. Risk can be defined quantitatively, for example, by a formula (probability × consequence) and so level of safety can be also defined quantitatively. However, "*Anshin*" (Assurance) can be defined as a state in that risk management is "assured (or trusted)" by the society, and thus it is oriented toward "process" rather than the level of safety itself.

I believe "safeguarding" public from risk is similar to the concept of "*Anshin*" (assurance). In other words, "safeguarding" would require "decision-making process" that public can be assured that they are protected from the risk. In short, issue of "safeguarding" the public is important as loss of public trust over nuclear safety has become a major issue in Japan.

Both the government and nuclear industry should make more sincere efforts in "safeguarding" the public from nuclear risk. How to recover public trust in Japan? One idea is to establish an independent oversight organization to serve public interests, rather than vested interest groups (Suzuki, 2019). Japan needs to establish such organization to gain public trust as establishing an independent NRA was not good enough.

---

7. Conclusion

Although the future of global nuclear industry is uncertain, it is clear that there are new challenges ahead in the area of safety, security, and safeguards. This chapter primarily discusses the challenges of security and safeguards. The major conclusions of based on the aforementioned analysis are as follows:

- Three major challenges exist for nuclear security: increased stockpile of fissile materials, dealing with new technologies such as cyberattacks, and institutional/organizational challenges dealing with internal threats.
- There are also three major challenges for safeguards: increasing number of new comers, dealing with noncompliance, and new type of requirements such as the one from JCPOA.
- To meet such challenges, IAEA must work with member governments and stakeholders while continuing to increase transparency of their operations.
- New technologies, such as information technologies (IT), may provide opportunities and/or pose potential new risks. It is imperative for the IAEA, and concerned governments must assess the impact of new technologies.
- Constant improvement in nuclear security and safeguards will be essential to improve confidence and effectiveness of security and safeguards. Leadership of

the IAEA and relevant government (especially with large nuclear programs) is key.

References

Baylon, C., December 18, 2014. Drones Are an Increasing Security Issues for the Nuclear Industry. Expert Comment for the Royal Institute of International Affairs. https://www.chathamhouse.org/expert/comment/drones-are-increasing-security-issue-nuclear-industry. (Accessed 5 March 2020).

Bunn, M., Sagan, S., 2017. Edit "Insider Threats". Cornell University, Ithaca.

Bunn, M., Roth, N., Tobey, W.H., January 2019. "Revitalizing Nuclear Security in an Era of Uncertainty", Project on Managing the Atom. Harvard Kennedy School, Belfer Center for Science and International Affairs. https://www.belfercenter.org/publication/revitalizing-nuclear-security-era-uncertainty. (Accessed 5 March 2020).

Findlay, T., October 2015. "Proliferation Alert! : The IAEA and Non-Compliance Reporting", Project on Managing the Atom. https://www.belfercenter.org/sites/default/files/legacy/files/Proliferation%20Alert.pdf. (Accessed 5 March 2020).

IAEA, 29 May 2015. Implementation of the NPT Safeguards Agreement and relevant provisions of Security Council resolutions in the Islamic Republic of Iran. Report of the Director Genral. https://www.iaea.org/sites/default/files/gov-2015-34.pdf.

IAEA, 2020. The Database on Nuclear Power Reactors (Power Reactor Information System: PRIS). https://pris.iaea.org/PRIS/home.aspx. (Accessed 5 March 2020).

IAEA, January 9, 2017. Communication Dated 22 December 2016 Received from the Permanent Mission of the United States of America Concerning a Joint Statement on Mitigating Insider Threats. INFCIRC/908. https://www.iaea.org/sites/default/files/publications/documents/infcircs/2017/infcirc908.pdf. (Accessed 5 March 2020).

IAEA, June 2016. IAEA Safeguards: Preventing the Spread of Nuclear Weapons. IAEA Bulletin. https://www.iaea.org/sites/default/files/bull572-june2016_0.pdf. (Accessed 5 March 2020).

International Atomic Energy Agency(IAEA), 2011. Nuclear Security Recommendations on Physical Protection of Nuclear Material and Nuclear Facilities. INFCIRC/225/Rev.5. https://www-pub.iaea.org/MTCD/publications/PDF/Pub1481_web.pdf. (Accessed 5 March 2020).

Japan Atomic Energy Relations Organization (JAERO), February 2019. Genshiryoku ni kannsuru seron chosa (Public Opinion Survey on Nuclear Energy Utilization). https://www.jaero.or.jp/data/01jigyou/pdf/tyousakenkyu30/r2018.pdf. (Accessed 5 March 2020).

Leverett, E., May 2019. Cyber Insurance for Civil Nuclear Facilities: Risks and Opportunities. Research Paper of the Royal Institute of International Affairs. https://www.chathamhouse.org/sites/default/files/CHHJ7127-Cyber-Insurance-Short-190507-WEB.pdf. (Accessed 5 March 2020).

Livingstone, D., Baylon, C., Brunt, R., October 5, 2015. Cyber Security at Civil Nuclear Facilities: Understanding the Risks. The Royal Institute of the International Affairs (Chatham House). https://www.chathamhouse.org/publication/cyber-security-civil-nuclear-facilities-understanding-risks. (Accessed 5 March 2020).

Nuclear Threat Initiative. References for Cyber Incidents at Nuclear Facilities. https://www.nti.org/analysis/tools/table/133/(accessed 5 March 2020).

Research Center for Nuclear Weapons Abolition, Nagasaki University (RECNA), December 2019. Global Stockpile of Nuclear Materials. http://www.recna.nagasaki-u.ac.jp/recna/en-fmdata/a-world-of-potential-bombs-fissle-material-inventory. (Accessed 5 March 2020).

Schneider, M., Antony, F., et al., September 27, 2018. World Nuclear Industry Status Report 2019. https://www.worldnuclearreport.org/-World-Nuclear-Industry-Status-Report-2019-.html. (Accessed 5 March 2020).

Suzuki, T., 2018. "Nuclear security policy of the Obama administration — its achievements and issues left behind: an interview with Laura Holgate". The Journal for Peace and Nuclear Disarmament. https://www.tandfonline.com/doi/full/10.1080/25751654.2018.1526283. (Accessed 5 March 2020).

Suzuki, T., February 2019. Nuclear Energy Policy after the Fukushima Nuclear Accident: An Analysis of 'Polarized Debate' in Japan. IntechOpen. https://www.intechopen.com/online-first/nuclear-energy-policy-after-the-fukushima-nuclear-accident-an-analysis-of-polarized-debate-in-japan. (Accessed 5 March 2020).

Van Dine, A., Assante, M., Page, S., 2016. Outpacing Cyber Threats: Priorities for Cybersecurity at Nuclear Facilities. https://media.nti.org/documents/NTI_CyberThreats__FINAL.pdf. (Accessed 5 March 2020).

Varjoranta, T., June 2016. Deputy Director general and head of safeguards department, "optimizing IAEA safeguards". In: IAEA Safeguards: Preventing the Spread of Nuclear Weapons. IAEA Bulletin. https://www.iaea.org/sites/default/files/bull572-june2016_0.pdf. (Accessed 5 March 2020).

The White House, 2015. The Iran Nuclear Deal: What You Need to Know about the JCPOA. https://obamawhitehouse.archives.gov/sites/default/files/docs/jcpoa_what_you_need_to_know.pdf.

Index

'*Note*: Page numbers followed by "f" indicate figures and "t" indicates tables.'

A

Aarhus Convention, 196–198
Absorbed dose, 50
Absorber
  boron, 25–26
  cadmium, 26
  europium, 26
  properties, 25, 25t
  reaction, 26
  rod reactor control system, 23, 24f
Actinides, 69
Advanced gas-cooled reactor (AGR), 36
Alloyed uranium metal, 16
ASEAN Network of Regulatory Bodies on Atomic
    Energy (ASEANTOM), 146–148
As low as reasonably practicable (ALARP),
    38–39
Atomic bombs
  Hiroshima, 84–85
  Nagasaki, 84–85
Atomic Energy Society of Japan (AESJ), 110

B

Boiling water reactor (BWR), 56
Boron, 25–26, 59
Brownfield, 61
Burnup, 57

C

Cadmium, 26
Carnot cycle, 3, 6
Chernobyl accident, 85
Civilian nuclear industry, 232–233
Civil liability
  international nuclear liability regimes
    Convention on Supplementary Compensation
      for Nuclear Damage (CSC), 215–219
    harmonized international legal regime,
      220
    legal principles, 214–215
    nonsovereign extralegal initiatives, 219
    origins, 211–213
  legal issues, 209–210
  nuclear new build countries, 220–225
    geographic distribution, 210–211
    new entrants, 211
    volume, 210

Civil nuclear energy
  advanced gas-cooled reactor (AGR), 36
  contracts for difference (CFD), 32–33
  generation IV nonelectrical opportunities, 36, 37f
  high-temperature test reactor (HTTR), 34
  low-carbon electricity, 35
  low enriched uranium (LEU), 31–32
  nonelectrical role, 31
  Nuclear Industry Research Advisory Board
    (NIRAB), 37–38
  Nuclear Non-Proliferation Treaty (NPT), 30–31
  pressurized water reactor (PWR), 30–31
  research reactors, 31
  United Kingdom, 33–34
  very-high-temperature reactor (VHTR), 36
Computer security incident response (CSIR), 234
Condensate storage tank (CST), 101–102
Construction materials
  aluminum and alloys, 18
  graphite, 19
  magnesium and alloys, 18–19
  parameters, 18, 18t
  requirements, 18
  stainless steel, 19
  zirconium and alloys, 19
Contracts for difference (CFD), 32–33
Convention on Supplementary Compensation for
    Nuclear Damage (CSC)
  Article III.2(a), 216
  Article V, 216
  Article XI(b), 216–217
  Contracting Party's national law, 217–218
  national compensation amount, 216
  supplementary compensation amount, 216
Coolants
  disadvantages, 22
  gases, 21
  liquid metal, 21
  requirements, 19–20
  water, 20–21
Cooperation, 176–177
Coulombic interactions, 49–50
Cyberattacks, 233

D

Decommissioning, 60–61, 77–78
Deferred dismantlement, 60

243

Directly ionizing particles, 49
Dry storage, 58
Dual nuclear power plant, 4, 4f

E
Edison effect, 9
Efficiency, 2f
   Carnot cycle, 3
   defined, 2
   heat conversion, 3
Electrode clouds, 8, 8f
Emergency diesel generators (EDGs), 100
Emergency planning zones (EPZs), 152–156
Emergency Preparedness and Response
      Information Management System
      (EPRIMS), 143–144
Emergency Preparedness Review (EPREV)
      missions, 163–164
Energy conversion methods
   direct conversion, 4–9
   efficiency, 2–3, 2f
   fission energy, 9–13
   heat transformation, 3–4
Entombment, 61
EPR-related IAEA safety standards, 142–144
Equivalent dose, 53
Espoo Convention, 194–196
European Union Radiological Data Exchange
      Platform (EURDEP), 159–160
Europium, 26
Event reporting, 157–162
Explosions, 105–107, 106f
Extended planning distance (EPD), 152–153
Extraordinary nuclear occurrence (ENO),
      220–221

F
Fast breeder reactor (FBR), 56
Fission products, 69
Fluorescence in situ hybridization (FISH), 88f–89f
Fuel
   alloyed uranium metal, 16
   carbide uranium and plutonium, 16
   dispersion fuel, 16
   element structures
      fuel rods, 14, 15f
      structural elements, 13, 13f
   fabrication, 56, 67, 68t
   parameters, 15
   plutonium metal, 16
   pure uranium metal, 15
   rods, 14, 15f

   uranium oxides and plutonium, 16, 17t
Fukushima Daiichi accidents, 1, 100–111

G
Gas cooled reactor (GCR), 10, 53–54
Gas-phase nuclear reactor, 10
Gen-III+ reactor, 127–128, 156
Global nuclear industry, 229–230
Government Investigation Committee, 117
Graphite, 19

H
Heat transformation
   dual nuclear power plant, 4, 4f
   steam power plant, 3, 4f
   three-loop-circuit nuclear power plant, 4, 5f
Heavy water reactor (HWR), 53–54
Highly enriched uranium (HEU), 231–232
High-pressure core injection (HPCI) system,
      100–101, 103f
High-temperature test reactor (HTTR), 34

I
IAEA Action Plan on Nuclear Safety, 2011,
      137–138
IAEA-centered regulatory framework, 139–145
Information sharing, 157–162
Ingestion and commodities planning distance
      (ICPD), 152–153
Inherently safe nuclear reactors, 137
Institute of Nuclear Power Operators (INPO),
      176–177
Institutionalize cybersecurity, 234
Integrated Regulatory Review Service (IRRS),
      143–144
Inter-Agency Committee on Radiological and
      Nuclear Emergencies (IACRNE),
      143–144
International Atomic Energy Agency (IAEA), 89,
      176–177, 211
International Federation of Red Cross and Red
      Crescent Societies, 146–148
International nuclear liability regimes
   Convention on Supplementary Compensation for
      Nuclear Damage (CSC)
      Article III.2(a), 216
      Article V, 216
      Article XI(b), 216–217
      Contracting Party's national law, 217–218
      national compensation amount, 216
      supplementary compensation amount, 216
   harmonized international legal regime, 220

legal principles, 214–215
nonsovereign extralegal initiatives, 219
origins, 211–213
International Radiation Monitoring Information
    System (IRMIS), 160–161
International regulatory responses, 149–165
International X-ray and Radium Protection
    Committee, 83–84
Ionizing radiation
    acute effects, 83–85
    biological effects, 83
    concepts, 82
    epigenetics changes, 90–91
    fluorescence in situ hybridization (FISH),
      88f–89f
    International Atomic Energy Agency (IAEA),
      89
    International X-ray and Radium Protection
      Committee, 83–84
    linear energy transfer (LET), 82
    linear no threshold (LNT) model, 82
    linear quadratic (LQ) equation, 83
    low dose effects, 86–87
      environmental radiation exposures, 87
      radiation workers and patients, 86–87
    overexposed cases, 84–85
      atomic bombs, Hiroshima, 84–85
      atomic bombs, Nagasaki, 84–85
      Chernobyl accident, 85
    radiation-induced chromosome alterations,
      87–89
    radiation-induced direct damage, 89
    radiation protection, 91
    single-strand breaks (SSBs), 83
    transgenerational effects, 90
    X-rays, 81
Isolation condenser (IC), 100–101

J
Japan Nuclear Technology Institute (JANTI), 110
Japan's new regulatory regime, 118–120, 119f
Joint Comprehensive Plan of Action (JCPOA), 40

K
Kashiwazaki-Kariwa (KK) site, 123–125

L
Light water graphite reactor (LWGR), 56
Linear energy transfer (LET), 82
Linear no threshold (LNT) model, 82
Linear quadratic (LQ) equation, 83
Low enriched uranium (LEU), 31–32

M
Magneto-hydrodynamic (MHD) method, 11f
    fast reactors, 10
    gas-cooled reactor, 10
    gas-phase nuclear reactor, 10
    operating principle, 9–10
    uranium hexafluoride (UF_6), 10–11
    working fluid, 10
Milling, 55, 66–67
Moderators, 22–23, 22t
Mothballing, 61
Multilateral environmental agreements (MEAs),
    193

N
National Academy of Sciences Study
    hazards, 112
    nuclear safety
      culture, 115
      regulations, 114
    off-site emergency responses, 114–115
    plant systems, 112–113
    resources, 113
    risk assessment, 113–114
    training, 113
National Diet Independent Investigation
    Commission, 116
Neutron shield, 59
New regulatory requirements, Japan
    earthquake protection, 124
    emergency preparedness, 124–125
    power supply, 123–124
    tsunami protection, 124
Nonionizing radiation, 49
Nonpower applications, civil nuclear energy,
    29–38
Nuclear emergency preparedness and response
    (EPR)
    ASEAN Network of Regulatory Bodies on
      Atomic Energy (ASEANTOM), 146–148
    bilateral arrangements, 165–167
    cross-border coordination, 149–152
    emergency planning zones (EPZs), 152–156
    Emergency Preparedness and Response Informa-
      tion Management System (EPRIMS),
      143–144
    Emergency Preparedness Review (EPREV)
      missions, 163–164
    EPR-related IAEA safety standards, 142–144
    European Union Radiological Data Exchange
      Platform (EURDEP), 159–160
    event reporting, 157–162

Nuclear emergency preparedness and response
(EPR) (*Continued*)
extended planning distance (EPD), 152−153
Gen-III + European power reactor, 156
global arrangements, 165−167
IAEA Action Plan on Nuclear Safety, 2011,
137−138
IAEA-centered regulatory framework, 139−145
incorporation by reference, 144−145
information sharing, 157−162
ingestion and commodities planning distance
(ICPD), 152−153
inherently safe nuclear reactors, 137
Integrated Regulatory Review Service (IRRS),
143−144
Inter-Agency Committee on Radiological and
Nuclear Emergencies (IACRNE), 143−144
International Federation of Red Cross and Red
Crescent Societies, 146−148
International Radiation Monitoring Information
System (IRMIS), 160−161
international regulatory responses, 149−165
normative pull, 144−145
recovery phase, 138−139
regional arrangements, 165−167
relevant international treaty instruments,
140−141
risk-exposed contracting party, 150−151
safety, 136
transboundary coordination/harmonization,
151−152
Unified System for Information Exchange in
Incidents and Emergencies (USIE),
160−161
urgent protective action planning zone (UPZ),
152−153
validation, 162−165
World Association of Nuclear Operators
(WANO), 146−148
Nuclear fuel cycle, 49
conversion, 55
decommissioning, 60−61
enrichment, 56
fuel fabrication, 56
milling, 55
mining, 54−55
radioactive materials, presence/release of
conversion, 67, 68t
enrichment, 67, 68t
fuel fabrication, 67, 68t
impacts, 62
milling, 66−67
mining, 63−66, 66t
normalized annual collective dose, 62−63,
65t
nuclear reactor operation, 67−72
radiation dose to workers, 62, 64t
radionuclides, 62, 63f
reprocessing, 72−73, 74t
radionuclides, 50, 51t−52t
reactor operation, 56−57
spent nuclear fuel
permanent disposal, 59−60
reprocessing, 59
storage, 57−58
transportation, 58−59
steps of, 53−54, 54f
uranium enrichment, 53−54
Nuclear Industry Research Advisory Board
(NIRAB), 37−38
Nuclear new build countries programs
China, 221−222
geographic distribution, 210−211, 212t
India, 222−223
international convergence, 225
new entrants, 211, 213t
Russia, 222
United Arab Emirates (UAE), 224−225
United Kingdom, 223−224
United States, 220−221
volume, 210, 210t
Nuclear Non-Proliferation Treaty (NPT), 30−31
Nuclear operators, 233
Nuclear plant severe accidents
Atomic Energy Society of Japan (AESJ), 110
condensate storage tank (CST), 101−102
emergency diesel generators (EDGs), 100
explosions, 105−107, 106f
Fukushima Daiichi accidents, 100−111
high-pressure core injection (HPCI) system,
100−101, 103f
investigative studies, 110−115
isolation condenser (IC), 100−101
Japan Nuclear Technology Institute (JANTI),
110
National Academy of Sciences Study
hazards, 112
nuclear safety culture, 115
nuclear safety regulations, 114
off-site emergency responses, 114−115
plant systems, 112−113
resources, 113
risk assessment, 113−114
training, 113

Nuclear Regulation Authority (NRA), 110
post-Fukushima regulations, 117—130
    Gen-III+ reactor, 127—128
    Japan's new regulatory regime, 118—120,
      119f
    Kashiwazaki-Kariwa (KK) site, 123—125
    new regulatory requirements, Japan, 120—125
    Nuclear Regulation Authority restart, 121t
    United States, 125—128
post-Fukushima regulatory body structure, Japan,
    115—117
    Government Investigation Committee, 117
    National Diet Independent Investigation
      Commission, 116
primary containment vessel (PCV), 101—102,
    103f
reactor core isolation cooling (RCIC) system,
    100—101, 104—105
spent fuel pool (SFP), 107
    accident cause, 109—110
    unit 1 explosion, 107
    unit 3 explosion, 108
    unit 4 explosion, 108—109
Tokyo Electric Power Company, 101, 105
top of active fuel (TAF), 102—103
unit 1 event progression, 101—102
unit 2 event progression, 104—105
Nuclear power reactors, 1, 71t
    absorber, 23—26
    actinides, 69
    activation products, 69
    Chernobyl nuclear power plant, 1
    classification, 26—27
    construction materials, 18—19
    coolant, 19—22
    decommissioning, 77—78
    dual nuclear power plant, 4, 4f
    fission products, 69
      noble gases, 70
    fuel, 14—16
    fuel element structures, 13—14
    Fukushima Daiichi nuclear power plant, 1
    liquid effluents, 70
    moderators, 22—23
    solid radioactive waste, 70—72
    steam power plant, 3, 4f
    three-loop-circuit nuclear power plant, 4, 5f
    Three Mile Island nuclear power plant, 1
    US National Academy of Sciences, 111
Nuclear Regulation Authority (NRA), 110, 121t
Nuclear rocket engines (NRE), 13f

high-quality, 11
rocket thrust, 12
working fluid, 11—12
Nuclear safeguards, 235—238
Nuclear security, 231—235
    threats, 39
Nuclear Threat Initiative (NTI), 233—234

O

Organisation for Economic Cooperation and
    Development (OECD), 211

P

Peltier effect, 5
Plutonium metal, 16
Post-Fukushima regulations, 117—130
    Gen-III+ reactor, 127—128
    Japan's new regulatory regime, 118—120,
      119f
    Kashiwazaki-Kariwa (KK) site, 123—125
    new regulatory requirements, Japan, 120—125
    Nuclear Regulation Authority restart, 121t
    United States, 125—128
Pressurized water reactor (PWR), 30—31, 56
Price-Anderson Act, 220
Primary containment vessel (PCV), 101—102,
    103f
Public trust in nuclear safety, 230—231
Pursue transformation, 234

Q

Q factor, 6—7, 7f

R

Radiation hazards
    absorbed dose, 50
    coulombic interactions, 49—50
    directly ionizing particles, 49
    electrically charged particles, 49
    equivalent dose, 53
    ionization, 49
    nonionizing radiation, 49
    nuclear fuel cycle, 49
    shielding, 50
    tissue weighting factor, 53
    uncharged particles, 49
Radiation-induced chromosome alterations,
    87—89
Radiation-induced direct damage, 89
Radiation protection, 91
Rankine cycle, 3

Reactor core isolation cooling (RCIC) system, 100–101, 104–105
Recovery phase, 138–139
Reduce complexity, 234
Refining, 55
Relevant international treaty instruments, 140–141
Reputational risk, 199–205
    host country nuclear regulatory authority, 202–203
    international agreements, 201
    international standards and practices, 200
    nuclear liability and insurance, 204–205
    public acceptance, 201–202
    sustainability, 203–204
    sustained government commitment, 201–202
Research reactors, 31
Richardson formula, 8
Risk-exposed contracting party, 150–151
^{222}Rn, 66–67

S
Safeguards, 39–41
    Comprehensive Nuclear Test-ban Treaty (CTBT), 191–193
    IAEA safeguards agreements, 189–190
    international conventions, 180–193
    Model Protocol Additional to the Agreement(s) between State(s) and the International Atomic Energy Agency for the Application of Safeguards, 190–191
    reputational risk analysis, 198–205
    Treaty on the Non-Proliferation of Nuclear Weapons (NPT), 188–193
Safety, 41–43
    Convention on Assistance in the Case of a Nuclear Accident or Radiological Emergency, 184–185
    Convention on Early Notification of a Nuclear Accident, 183–184
    Convention on Nuclear Safety (CNS), 180–182
    international conventions, 180–193
    Joint Convention on the Safety of Spent Fuel Management and on the Safety of Radioactive Waste Management, 182–183
    reputational risk analysis, 198–205
Safety assessment principles (SAPs), 38–39
Security, 38–39
    amendment, 186–187
    Convention on the Physical Protection of Nuclear Material (CPPNM), 185–186

International Convention for the Suppression of Acts of Nuclear Terrorism, 187–188
    international conventions, 180–193
    reputational risk analysis, 198–205
Seebeck effect, 5
Shielding, 50
    layers of, 59
    neutron, 59
Single-strand breaks (SSBs), 83
Solid radioactive waste, 70–72
Solvent extraction, 55
Special Drawing Rights (SDRs), 214–215
Spent fuel pool (SFP), 107
    accident cause, 109–110
    unit 1 explosion, 107
    unit 3 explosion, 108
    unit 4 explosion, 108–109
Spent nuclear fuels, 57
    permanent disposal, 59–60
    radionuclides, 73–75
    reprocessing, 59
    storage, 57–58
    transportation, 58–59
Stainless steel, 19
Steam power plant, 3, 4f

T
Thermal efficiency, 3
Thermal electron energy converter (TEC), 7–8
Thermionic emission, 8
Thermoelectric generators (TEG), 6f
    Carnot cycle, 6
    Peltier effect, 5
    Q factor, 6–7, 7f
    Seebeck effect, 5
Three-loop-circuit nuclear power plant, 4, 5f
Three Mile Island nuclear power plant, 1
Tissue weighting factor, 53
Tokyo Electric Power Company, 101, 105
Top of active fuel (TAF), 102–103
Transgenerational effects, 90
Tributyl phosphate (TBP), 55

U
Unified System for Information Exchange in Incidents and Emergencies (USIE), 160–161
Unit 1 event progression, 101–102
Unit 2 event progression, 104–105
Uranium enrichment, 53–54
Uranium hexafluoride (UF$_6$), 10–11, 53–54
    enrichment, 56

Uranium mining, 63–66, 66t
  open-pit (surface), 54–55
  in situ leaching (ISL), 54–55
  underground mining, 54–55
Urgent protective action planning zone (UPZ),
  152–153
US National Academy of Sciences, 111
US Nuclear Regulatory Commission (NRC),
  220

V

Very-high-temperature reactor (VHTR), 36
Voltage, 8–9, 8f–9f

W

Wet storage, 58
World Association of Nuclear Operators (WANO),
  146–148, 176–177

X

X-rays, 81

Y

Yellowcake (U_3O_8), 55

Z

Zirconium, 19